THE HANDBOOK OF METEOROLOGY

Frank R. Spellman

SCARECROW PRESS, INC.
Lanham • Toronto • Plymouth, UK
2013

Published by Scarecrow Press, Inc.
A wholly owned subsidiary of The Rowman & Littlefield Publishing Group, Inc.
4501 Forbes Boulevard, Suite 200, Lanham, Maryland 20706
www.rowman.com

10 Thornbury Road, Plymouth PL6 7PP, United Kingdom

British Library Cataloguing in Publication Information Available

Library of Congress Cataloging-in-Publication Data

Spellman, Frank R.
  The handbook of meteorology / Frank R. Spellman.
     p. cm.
  Includes bibliographical references and index.
  Summary: "Book offers a condensed yet comprehensive survey of the science of weather: temperature, pressure, humidity, wind, pressure systems, fronts, storms, weather forecasts, cloud formation, weather tools, etc., with tables, a glossary, and illustrations to translate detailed technical information into terms that everyone can follow"—Provided by publisher.
  ISBN 978-0-8108-8612-4 (cloth : alk. paper) — ISBN 978-0-8108-8613-1 (ebook)
  1. Meteorology—Handbooks, manuals, etc. I. Title.
  QC864.S64 2013
  551.5—dc23   2012027997

♾™ The paper used in this publication meets the minimum requirements of American National Standard for Information Sciences—Permanence of Paper for Printed Library Materials, ANSI/NISO Z39.48-1992.

Printed in the United States of America

For Nancy Velasquez

Contents

Preface   ix

Part One: Introduction   1

1   Setting the Stage   2
    U.S. National Weather Service   3
    Weather and Climate   14
    Meteorology: The Science of Weather   15
    Wind and Breezes   16
    Air Masses   18
    Thermal Inversions and Air Pollution   18
    References and Recommended Reading   19

Part Two: Meteorological Basics   21

2   Weather Elements   22
    Air Pressure   23
    Precipitation   25
    Wind   27
    Air Temperature   29
    Relative Humidity   31
    Solar Energy   32
    Evaporation   33
    Pollution   33
    Cloud Heights   33

Gas Laws   34
Boyle's Law   34
Charles's Law   35
Ideal Gas Law   36
Flow Rate   38
Gas Conversions   38
Major Constituents   39
Both Major and Minor Constituents   39
Minor Constituents   40
Gas Density   40
Heat Capacity and Enthalpy   41
Heat and Energy in the Atmosphere   41
Adiabatic Lapse Rate   41
Viscosity   42
References and Recommended Reading   43
3 The Atmosphere   44
Earth's Thin Skin   47
The Troposphere   48
The Stratosphere   49
A Jekyll-and-Hyde View of the Atmosphere   49
Atmospheric Particulate Matter   50
References and Recommended Reading   51
4 Moisture in the Atmosphere   52
Cloud Formation   53
Major Cloud Types   54
Moisture in the Atmosphere   56
References and Recommended Reading   57
5 Precipitation and Evapotranspiration   58
Phases of Precipitation   60
Precipitation   62
Coalescence   62
Bergeron Process   63
Types of Precipitation   64
Convectional Precipitation   64
Orographic Precipitation   64
Frontal Precipitation   65
Evapotranspiration   65
Evaporation   65
Transpiration   66
The Process of Evapotranspiration   66
References and Recommended Reading   67

6   The Seasons   68
    Why Do We Have the Seasons?   69
    Why Are Sunrise and Sunset Not Exactly 12 Hours Apart
        at Equinox?   71
    Heat Island Effect in Large Cities   71
    Anthropogenic (Human-Caused) Effects on Precipitation   73
    References and Recommended Reading   75
7   Earth's Radiation Budget   77
    Energy   78
    Potential Energy   78
    Kinetic Energy   78
    Thermal Properties   80
    Specific Heat   81
    Transferring Energy   81
    Earth's Heat Balance   83
    Earth's Radiation Budget   83
    Reradiation   84
    Insolation   86
    Solar Constant   86
    Transparency   87
    Daylight Duration   87
    Angle of Sun's Ray   88
    Heat Distribution   88
    Differential Heating   88
    Transport of Heat   89
    Global Distribution of Heat   89
    Albedo   90
    Scattering   91
    Absorption   92
    Climate Change and Greenhouse Effect   92
    The Past   94
    Greenhouse Effect   100
    References and Recommended Reading   104

Part Three: Weather and Climate   105

8   The Science of Weather and Climate   106
    Meteorology: The Science of Weather   107
    El Niño / Southern Oscillation   108
    Air Masses   109
    Optical Phenomena   113
    Fata Morgana   115
    References and Recommended Reading   116

9   Microclimates   118
Types of Microclimates   119
Microclimates Near the Ground   120
Microclimates Over Open Land Areas   120
Microclimates in Woodlands or Forested Areas   121
Microclimates in Valley and Hillside Regions   121
Microclimates in Urban Areas   122
Microclimates in Seaside Locations   122
References and Recommended Reading   123

Part Four: Atmospheric Dynamics   **125**

10   The Atmosphere in Motion   126
Global Air Movement   127
Earth's Atmosphere in Motion   128
Causes of Air Motion   128
Local and World Air Circulation   133
Blowing in the Wind . . .   136
References and Recommended Reading   138
11   Weather Forecasting Tools   139
Butterfly Effect   139
Predicting Weather   140
Meteorological Tools   140
Wind Speed   141
Wind Direction   145
Temperature and Temperature Difference   147
Radiation   148
Mixing Height   149
System Performance   150
Quality Assurance and Quality Control   152
References and Recommended Reading   154

Glossary   155

Index   217

About the Author   223

Preface

WANT TO HAVE A SAFE CONVERSATION? Do you need to break the ice (start a conversation) with someone of interest to you? Or, do you simply want to start small talk with anyone, including a total stranger? Hey, there is no better way to do all the above than by beginning your conversation, small talk or even more, with a discussion about the weather. You know, we simply state to everyone or anyone who is listening that it is snowing, raining, or sunny outside, and then we ask, "How about you? What do you think?" Or, in the case of long distance communication, you might ask, "What's the weather like in your neck of the woods?"

Weather not only is a common way in which we begin a conversation but also is common verbiage seen on the Twitter website. You know, Twitter, the SMS (short message service) that we all (sooner or later) use to communicate with those we wish to (and some we do not wish to). We simply post our text-based tweets and refer to or ask about the weather, and then get on with it—on to those areas we really want to talk about.

Then there are those folks who have one particular passion that sustains them in the outdoors, or maybe several outdoor activities. For some, it's the experience of carrying a backpack through an alpine meadow sprinkled with a host of colors of countless wildflowers. For others, it's the vision of fly fishing for trout on a clear river perfectly mirroring the surrounding tree-lined mountain landscape. Some look for snow, for winter activities. Still others are in their element, their glory, their habitat(s) where wind and leaf-fall drop quaking aspen leaves that look like golden doubloons in the sunlight. Still others just like and crave the stillness of a frosty cold walk in the woods. A few of these nature lovers enjoy this same walk but prefer the silence broken by the chirp of a solitaire bird or the bugling of a bull elk or the sighing of the wind through tall pines. Now, how enjoyable are any of these pursuits in a heavy rain storm, a blizzard, a typhoon, a hurricane, or a tsunami?

The fact is before we hike, rock climb, jet ski, sky dive, mountain bike, hot air balloon, canoe across a lake, hike the Kaibab Trail in North Rim of Grand Canyon or climb to the summit of Angel's Landing in Zion National Park, or cross-country ski anywhere, we usually check the weather forecast—yet, even so, almost everyone has been caught out in a storm unexpectedly. In *The Handbook of Meteorology*, it is my intention to give you the basic tools you need to stay out of the storm's path when you're headed out for a long hike or even an afternoon jog around the pond, lake, or city park. One of the keys to staying comfortable and safe in the outdoors is the ability to recognize the signs that indicate that weather conditions may be about to make a dramatic change.

In addition to providing basic meteorological information to those who might want to pursue outdoor activities, and thus be able to plan ahead, *The Handbook of Meteorology* provides other weather questions that we all have and want answered. These questions, based on personal experience, are the easiest to ask but the most difficult to answer. If a process is essential to a true understanding of the basic dynamics of weather on earth, then it is included in this text, along with the best, plain English, condensed, non-gibberish (avoiding all forms of gobbledygook) explanation science has to offer.

The Handbook of Meteorology has been compiled to give specialists, non-specialists, and generalists a real understanding of the way our weather functions. It provides scientific (natural) answers to questions that arise when looking at or studying the world around us—answers that strip away the mystery and cut through to the basic foundational processes.

This handbook is for people who want to refresh their memories and/or learn about meteorology without taking a formal course. From weather basics—temperature, pressure, humidity, and wind; highs, lows, fronts, and storms; weather forecasts, cloud watching, weather tools—and much more, *The Handbook of Meteorology* provides a condensed but all-inclusive broad sweep of meteorology, employing several illustrations to translate detailed technical information into terms that everyone can follow and readily reference. It can serve as a classroom supplement, tutorial aid, self-teaching guide, home-schooling reference, and comprehensive desk reference for any budding meteorologist or environmental professional in the field or laboratory.

Based on personal experience, I have found that many professionals do have some background in meteorology (Don't we all? How can anyone avoid the weather?), but many of these folks need to stay current (remember that there is nothing static about nature or science—and the same can be said, hopefully, about our understanding of meteorology).

The Handbook of Meteorology fills the gap between general introductory texts and the more advanced environmental science books used in graduate courses. The handbook fills this gap by surveying and covering the basics of natural sciences and studies. This book is both a technical and nontechnical survey for those with or without a background in meteorological studies—presented in a reader-friendly written style.

The study of meteorology is multidisciplinary in that it incorporates aspects of biology, chemistry, physics, ecology, geology, nature, pedology, sociology, and many other fields. Books on the subject are typically geared toward professionals in these fields. This makes undertaking a study of meteorology daunting to those without this specific background. However, this complexity also indicates meteorology study's broad scope of impact. Because weather affects us, sometimes in profound double-edged ways—one edge presents incomparable aesthetics and beauty and the other edge presents such events as floods, tornadoes, hurricanes, typhoons, and so forth—it is important to understand at least some of the basic concepts of the discipline.

Along with basic natural scientific principles, the handbook provides a clear, concise presentation of the consequences of the anthropogenic interactions with the environment we inhabit. Even if you are not tied to a desk, the handbook provides you, the naturalist or non-naturalist, with the jargon, concepts, and key concerns of meteorology and meteorology in action.

This reference book is compiled in an accessible, user-friendly format, unique in that it explains scientific concepts in the most basic way possible. Organized in dynamic themes that reflect the approach of modern science, *The Handbook of Meteorology* draws together common processes and activities rather than dividing and classifying. The beginning weather student will quickly have a foundation for the future. The more experienced may find their knowledge suddenly snapped into focus with a new and solid foundation under that which is already known. And remember, weather is a topic that begins many conversations—breaks the ice, so to speak—thus, having knowledge of weather dynamics can make any conversation more enlightening to all involved.

Unless otherwise noted, discussions included throughout this book pertain to weather in the northern hemisphere, more specifically to the United States and Canada.

Frank R. Spellman
Norfolk, VA

I

INTRODUCTION

I bring fresh showers for the thirsting flowers,
From the seas and the streams;
I bear light shade for the leaves when laid
In their noon-day dreams.

—Percy Bysshe Shelley, *The Cloud*, 1820

1

Setting the Stage

Just as there are people with distorted, failing, or nonexistent senses of smell, there are those at the other end of the olfactory spectrum, prodigies of the nose, the most famous of whom is probably Helen Keller. "The sense of smell," she wrote, "has told me of a coming storm hours before there was any sign of it visible. I notice first a throb of expectancy, a slight quiver, a concentration in my nostrils. As the storm draws near my nostrils dilate, the better to receive the flood of earth odors which seem to multiply and extend, until I feel the splash of rain against my cheek. As the tempest departs, receding farther and farther, the odors fade, become fainter and fainter, and die away beyond the bar of space." Other individuals have been able to smell changes in the weather, too, and, of course, animals are great meteorologists (cows, for example, lie down before a storm). Moistening, misting, and heaving the earth breathes like a great dark beast. When barometric pressure is high, the earth holds its breath and vapors lodge in the loose packing and random crannies of the soil, only to float out again when the pressure is low and the earth exhales. The keen-nosed, like Helen Keller, smell the vapors rising from the soil, and know by that signal that there will be rain or snow. This may also be, in part, how farm animals anticipate earthquakes—by smelling ions escaped from the earth. (Ackerman, 1990, p. 44–45)

Information in this chapter from NOAA's National Weather Service Public Affairs Office: Evolution of the National Weather Service (2010). Accessed May 20, 2011, at www.nws.noaa.gov/pa/history/timeline.php.

ONE OF THE PRIMARY REASONS PEOPLE are so intensely interested in weather is our lack of control over it. As mentioned in the preface, we are intimately affected by weather in our lives. We dress as we do because of it. We plan our lives around it. We work because of it—shoveling snow, raking leaves, planting gardens around it. We go places because of it, or move away from it. We make it the first topic of conversation after "How do you do?" We joke about it (If you don't like the weather in [your town's name here] just stick around—it'll change). But we can't change it. And it can change us.

This, of course, sometimes terrifies us, chiefly because in its extremes of temperature, velocity, and precipitation, we know that weather can destroy us. Tornadoes, hurricanes, too much or too little rain, heavy snows, extreme cold, and extreme heat damage and kill. All we can do is take shelter from the adverse conditions and rebuild when they pass, or plant our crops again and hope for better.

The only important way that we can affect our weather is one harmful to us—we can pump pollutants into our air, so that even un-extreme weather can affect us adversely as well.

U.S. National Weather Service

In regard to the evolution of the National Weather Service (1849–1994), NOAA (2010) points out that the service has its beginning in the early history of the United States. Weather always has been important to the citizenry of this country, and this was especially true during the 17th and 18th centuries. Weather also was important to many of the Founding Fathers. Colonial leaders who forged the path to our country's independence were also avid weather observers. Thomas Jefferson bought his first thermometer while writing the Declaration of Independence and purchased his first barometer a few days following the signing of the document. Incidentally, he noted that the high temperature in Philadelphia, Pennsylvania, on July 4, 1776, was 76°. Jefferson made regular observations at Monticello for 1772–1778 and participated in taking the first known simultaneous weather observations in America. George Washington also took regular observations; the last weather entry in his diary was the day before he died.

During the early and mid 1800s, weather observation networks began to grow and expand across the United States. Although most basic meteorological instruments had existed for over 100 years, it was the telegraph that was largely responsible for the advancement of operational meteorology during the 19th century. With the advent of the telegraph, weather observations from distant points could be "rapidly" collected, plotted, and analyzed at one location. The following presents a brief chronological history of meteorological technical development and the U.S. Weather Service.

1849: Smithsonian Institution supplies weather instruments to telegraph companies and establishes an extensive observation network. Observations submitted by telegraph to the Smithsonian, where weather maps are created. By the end of 1849, 150 volunteers throughout the United States were reporting weather observations to the Smithsonian regularly. By 1860, 500 stations were furnishing daily telegraphic weather reports to the *Washington Evening Star*, and as the network grew, other existing systems were gradually absorbed, including several state weather services.

1860: Five hundred stations are making regular observations, but work is interrupted by the Civil War.

1869: Telegraph service, instituted in Cincinnati, begins collecting weather data and producing weather charts. The ability to observe and display simultaneously observed weather data through the use of the telegraph quickly led to initial efforts toward the next logical advancement, the forecasting of weather. However, the ability to observe and forecast weather over much of the country required considerable structure and organization—a government agency.

1870: A joint congressional resolution requiring the secretary of war "to provide for taking meteorological observations at the military stations in the interior of the continent and at other points in the States and Territories . . . and for giving notice on the northern (Great) Lakes and on the seacoast by magnetic telegraph and marine signals, of the approach and force of storms" was introduced. The resolution was passed by Congress and signed into law on February 9, 1870, by President Ulysses S. Grant. An agency had been born that would affect the daily lives of most of the citizens of the United States through forecasts and warnings.

1870–1880: Ben. Albert J. Meyer serves as the first director of the Weather Bureau. Gen. William Babcock serves as the director of the Weather Bureau.

1887–1891: Maj. Gen. Adolphus Greely takes over as director of the Weather Bureau.

May 30, 1889: An earthen dam breaks near Johnstown, Pennsylvania. The flood kills 2,209 people and wrecks 1,880 homes and businesses.

October 1, 1890: Weather Service is first identified as civilian enterprise when Congress, at the request of President Benjamin Harrison, passes an act creating the Weather Bureau in the Department of Agriculture.

1891: The secretary of agriculture directs R. G. Dyrenforth to carry out rainmaking experiments by setting off explosions from balloons in the air; Weather Bureau becomes responsible for issuing flood warnings to the public; telegraphic reports of stages of rivers were made at 26 places on the Mississippi and its tributaries, the Savannah and Potomac Rivers.

1891–1895: Professor Mark W. Harrington replaces Maj. Gen. Greely as director of the Weather Bureau.

1894: William Eddy, using five kites to loft a self-recording thermometer, makes first observations of temperatures aloft.

September 30, 1895: The first daily Washington weather map is published by the Weather Bureau.

1895–1913: Secretary of Agriculture J. Sterling Morton appoints Professor Willis Luther Moore chief of the Weather Bureau.

1898: President William McKinley orders the Weather Bureau to establish a hurricane warning network in the West Indies.

1900: Cable exchange of weather warnings and other weather information begins with Europe.

September, 1900: Hurricane strikes Galveston, Texas, killing over 6,000 people. The wife of the Galveston official in charge and one Weather Bureau employee and his wife are killed in the associated flooding. Weather Bureau forecast the storm four days earlier, but not the high tide.

1901: Official three-day forecasts begin for the North Atlantic. At the Weather Bureau conference in Milwaukee, Wisconsin, Willis Moore observes the post office department is delivering slips of paper, on which are daily forecasts frost and cold-wave warnings, to everyone's door with the mail. The one disadvantage to the system is the mail carriers start their routes about 7 a.m. and that day's forecast is not issued until 10 a.m., so the previous night's forecasts are used.

1902: The Marconi Company begins broadcasting Weather Bureau forecasts by wireless telegraphy to Cunard Line steamers; the Bureau begins collecting flood damage statistics nationally.

1903: Weather sensitive historic events: United States and Panama sign the Canal Treaty; the first automobile trip across the United States is completed from San Francisco to New York City; Orville Wright makes first powered airplane flight at Kill Devil Hill, North Carolina, after consultation with the Weather Bureau.

1904: Government begins using airplanes to conduct upper air atmospheric research.

1905: The SS *New York* transmits the first wireless weather report received on ship at sea.

1907: Weather sensitive historic event: Round-the-world cruise of U.S. "Great White Fleet" including 16 battleships and 12,000 men.

1909: The Weather Bureau begins its program of free-rising balloon observations.

1910: Weather Bureau begins issuing generalized weekly forecasts for agricultural planning; the River and Flood Division begins assessment of water available each season for irrigating the Far West.

1911: The first transcontinental airplane flight, from New York City to Pasadena, California, by C. P. Rogers, in 87 hours and 4 minutes, air time, over a period of 18 days.

1912: As a result of the Titanic disaster, an international ice patrol is established, conducted by the Coast Guard; first fire weather forecast issued; Dr. Charles F. Marvin appointed chief of the Weather Bureau.

1913–1934: Professor Charles F. Marvin serves as the new chief of the Weather Bureau, replacing Professor Moore.

1914: An aerological section is established within the Weather Bureau to meet growing needs of aviation; first daily radiotelegraphy broadcast of agricultural forecasts by the University of North Dakota.

1916: Fire Weather Service established, with all district forecast centers authorized to issue fire weather forecasts; the Bureau's fire district forecast center started at Medford, Oregon.

1917: Norwegian meteorologists begin experimenting with air mass analysis techniques that will revolutionize the practice of meteorology.

Did You Know?

The weather map is the most valuable tool that the meteorologist uses to forecast the weather. Without this tool, it would be very hard to predict what the weather was going to do. Weather maps summarize what is happening in the atmosphere at a certain time, and it would be very difficult to predict changes to the weather without these maps. By looking at weather maps from different heights in the atmosphere, a meteorologist can make a three-dimensional picture of what is happening in the atmosphere. They can tell whether a particular area has high or low pressure, whether it may rain, and many other things just by looking at a weather map (NOAA, 2007).

1918: The Weather Bureau begins issuing bulletins and forecasts for domestic military flights and for new air mail routes.

1919: Navy Aerological Service established on a permanent basis; first transatlantic flight by U.S. Navy sea plane, with stops in Newfoundland, Azores, and Lisbon.

1920: Meteorologists form a professional organization, the American Meteorological Society.

1921: The University of Wisconsin makes a radiotelephone broadcast of weather forecasts, the first successful use of the new medium for weather advisories.

1922: Histories of 500 river stations completed.

1926: The Air Commerce Act directs the Weather Bureau to provide for weather services to civilian aviation; fire weather service formally inaugurated when Congress provides funds for seven fire weather districts.

1927: The Weather Bureau establishes a West Coast prototype for an Airways Meteorological Service; Charles A. Lindbergh flies alone from Long Island, nonstop, to Paris. The 3,610-mile trip is completed in 33.5 hours. As on his earlier transcontinental flight, he consulted the bureau in planning this flight. However, Lindbergh didn't wait for the final confirmation of good weather over the Atlantic. When Weather Bureau officials in New York heard that Lindbergh had left, they expressed surprise because the forecasts indicated the flight should have been delayed by at least 12 hours. Indeed, Lindbergh ran into problems with fog and rain—as the Weather Bureau had predicted.

1928: The teletype replaces telegraph and telephone service as the primary method for communicating weather information.

1931: The Weather Bureau begins regular 5 a.m. EST aircraft observations at Chicago, Cleveland, Dallas, and Omaha, at altitudes reaching 16,000 feet. This program spells the demise of the "kite stations."

1933: A science advisory group apprises President Franklin D. Roosevelt that the work of the volunteer cooperative earth observer network is one of the most extraordinary services ever developed, netting the public more per dollar expended than any other government service in the world. By 1990 the 25-mile radius network encompasses nearly 10,000 stations.

1934: Bureau establishes an Air Mass Analysis Section.

1934–1937: "Dust Bowl" drought in southern plains causes severe economic damage.

1934–1938: Dr. Willis L. Gregg is named bureau chief, replacing Professor Marvin.

1935: A hurricane warning service is established; the Smithsonian Institution begins making long-range weather forecasts based on solar cycles; floating automatic weather instruments mounted on buoys begin collecting marine weather data.

1936: Hoover Dam is completed, a weather sensitive engineering feat.

1937: First official Weather Bureau radio meteorograph, or radiosonde, sounding made at East Boston, Massachusetts. This program spells the end for aircraft soundings since balloons average only 50,000 feet altitude. Twelve pilots die flying weather missions. January flood on the Ohio River is the greatest ever experienced, with the Ohio River levels exceeding all previous. Cincinnati's 80-foot crest and Louisville's 81.4-foot crest have never been exceeded. Seventy percent of Louisville is under water; 175,000 of its residents flee their homes; the entire city of Paducah, Kentucky—population 40,000—is evacuated.

1938–1963: President Franklin D. Roosevelt appoints Dr. Francis W. Reichelderfer chief of the Weather Bureau.

Did You Know?

A meteorologist must convey a lot of information without using a lot of
words. When looking at a weather map, a meteorologist needs to know
where the cold air is, where the warm air is, where it is raining, what type of
clouds are in the area, and many more things. The reason for this is that forecasts need to be accurate. But, they also need to be timely (NOAA, 2007).

1939: Bureau initiates automatic telephone weather service in New York City;
radio meteorographs, or radiosondes, replace all military and Weather
Bureau aircraft observations.
1940: Weather Bureau transferred to Department of Commerce; Army and
Navy establish weather centers; President Roosevelt orders Coast Guard to
man ocean weather stations.
1941: Dr. Helmut Landsberg, "the Father of Climatology," writes the first edi-
tion of his elementary textbook, titled *Physical Climatology*. Two women
are listed as observer and forecaster in the Weather Bureau.
1942: A Central Analysis Center, forerunner of the National Meteorological
Center, is created to prepare and distribute master analyses of the upper
atmosphere; Joint Chiefs of Staff establish a Joint Meteorological Com-
mittee to coordinate wartime civilian and military weather activities; Navy
gives the Weather Bureau 25 surplus aircraft radars to be modified for
ground meteorological use, marking the start of a weather radar system
in the United States. Navy aerologists play key role as U.S., carrier-based
Navy planes decimate Japanese fleet in mid-Pacific Battle of Midway Island
in early June 1942, a turning point in World War II. A cooperative thunderstorm research effort is undertaken by the bureau, military services, and
the University of Chicago.
1944: The decision to invade Normandy on June 6th was based on weather
forecasts, which indicated the correct combination of tides and winds.

Did You Know?

If too much time is spent making the forecast, it will be late. Not many
people want to know what the weather was doing 20 minutes ago. Most
people want to know what the weather is going to do in the near future.
Because of this, weather symbols were invented so that weather maps
could be looked at in a short amount of time (NOAA, 2007).

1945: Over 900 women are employed by the Weather Bureau as observers and forecasters, as a result of filling the positions of men during World War II.

1946: The U.S. Weather Bureau selects Cincinnati, Ohio, and Kansas City, Missouri, as locations for the nation's first hydrologist-staffed River Forecast Centers. Eventually, 13 RFC's would be established to serve the United States.

1948: USAF Air Weather Service meteorologists issue first tornado warnings from a military installation. Princeton's Institute for Advanced Studies begins research into use of a computer for weather forecasting; Chicago Weather Bureau office demonstrates use of facsimile for map transmission; truck mounted campers first used as mobile forecast stations in major forest fires.

1950: The Weather Bureau begins issuing 30-day weather outlooks; authorizes release of "tornado alerts" to the public.

1951: Severe Weather Warning center begins operation at Tinker Air Force Base, Oklahoma, forerunner of the National Severe Storms Center; World Meteorological Organization established by the United Nations. Bureau Chief Reichelderfer elected its first head; bureau's New Orleans data tabulation unit moves to Asheville, North Carolina, to become the National Weather Records Center and later the National Climatic Data Center.

1952: Bureau organizes Severe Local Storms forecasting unit in Washington, DC, and begins issuing tornado forecasts.

1954: The Weather Bureau, Navy, Air Force, MIT's Institute for Advanced Study, and University of Chicago form a Joint Numerical Weather Prediction Unit at Suitland, Maryland. This will become a twice daily routine in 1955, using an IBM 701. First radar specifically designed for meteorological use, the AN/CPS-9, is unveiled by the Air Weather Service, USAF.

1955: Hurricane Diane floods the Northeast and 187 people die. Regularly scheduled operational computer forecasts begun by the Joint Numerical Forecast Unit. Weather Bureau becomes a pioneer civilian user of computers along with the Census Bureau in Commerce; bureau begins development of Barotropic model, a first for numerical predictions.

1956: The bureau initiates a National Hurricane Research Project.

1957–1958: International Geophysical year provides first concerted worldwide sharing of meteorological research data. Dr. Reichelderfer accepts proposal by Dr. James Brantly of Cornell Aeronautical Laboratories to modify surplus Navy Doppler radars for several storms' observation—the first endeavor to measure motion of precipitation particles by radar.

1958: Weather-related scientific event: Explorer I is launched into space by an Army Redstone Rocket from Cape Canaveral. This satellite discovers the Van Allen radiation belt; the National Meteorological Center is established; the first commercial jet passenger flight from New York to Miami by National Airlines.

1959: Major weather scientific event: The Army launches Vanguard II from Cape Canaveral, carrying two photocell units to measure sunlight reflected from clouds, demonstrating feasibility for a weather satellite. The bureau's first WSR-57 weather surveillance radar is commissioned at the Miami hurricane forecast center. Same model, now obsolete, is still in service in New York City, although replacement parts must be machined by hand; The Naval Aerological service becomes the Naval Weather Service.

The Thomas Jefferson and John Campanius Holm awards are created by the Weather Bureau to honor volunteer observers for unusual and outstanding accomplishments in the field of meteorological observations.

1960: World's first weather satellite, solar orbiting TIROS I, successfully launches for the Air Force Missile Test Center at Cape Canaveral, Florida; the Weather Bureau and NASA invite scientists from 21 nations to participate in the analysis of weather data gathered by TIROS II. In cooperation with the Department of Health, Education and Welfare, Weather Bureau meteorologists issue first advisories on air pollution potential over the Eastern United States.

1961: President Kennedy, in his State of the Union address, invites all nations to join the United States in developing an International Weather Protection Program. The Weather Bureau assumes full responsibility for severe weather forecasting, establishing the National Severe Storms Center in Kansas City; special training begins for Federal Aviation Authority employees to equip them to brief pilots as part of a joint FAA-Bureau program; the USAF Air Weather Service issues first official forecast of clear air turbulence; scientists from 27 countries attend NASA Weather Bureau–sponsored internal workshop on technique to interpret weather satellite data.

1963–1965: Dr. Robert M. White succeeds Dr. Reichelderfer as chief of the Weather Bureau.

Did You Know?

Because a large part of the United States was not well populated in the late 1800s and early 1900s, the weather maps of the day were missing observations from most of the central plains states. Also, meteorologists did not always understand what the maps they were analyzing meant. Many of the theories used today in forecasting had not yet been developed (NOAA, 2007).

1963: Commerce Department polar-orbiting weather satellite TIROS III is launched with automatic picture transmission (APT) capability, eventually to provide continuous cloud images to over 100 nations.

1964: Secretary of Commerce establishes the office for the Federal Coordinator for meteorology; the National Severe Storms Laboratory is established in Norman, Oklahoma; the American Meteorological Society writes to the Taiwanese Ambassador to the United States deploring treatment accorded Mr. Kenneth T. C. Cheng, head of the Taiwan Weather Service, who had been indicted for an incorrect typhoon forecast. The AMS points out that if forecasters were indicted for an incorrect forecast there could soon be a total lack of forecasters. (Minutes of the AMS Council, October 3–4, 1964.)

1965: Environmental Science Services Administration is created in the Department of Commerce, incorporating the Weather Bureau and several other agencies; Dr. White is appointed administrator.

1965–1979: Dr. George Cressman is named Weather Bureau director.

1966: Weather officials from 25 nations meet in London for the First International Clean Air Congress; National Meteorological Center introduces a computer numerical model capable of making sea level predictions as accurate as those made manually.

1967: Responsibility for issuing air pollution advisories is assigned to the Weather Bureau; National Meteorological Center fire weather forecasts extended to cover contiguous United States.

1969: Weather-related historic event: Neil Armstrong, commander of spacecraft Apollo 11, becomes first man to set foot on the moon.

1970: Environmental Science Services Administration (ESSA) becomes the National Oceanic and Atmospheric Administration (NOAA), with Dr. White as administrator. U.S. Weather becomes National Weather Service.

1972: Rainfall from Hurricane Agnes floods the east coast killing 105 people; a flash flood in the Black Hills of South Dakota kills 237 people.

1973: National Weather Service purchases its second generation radar (WSR-74).

1975: The first "hurricane hunter" Geostationary Operational Environmental Satellite (GOES) is launched into orbit; these satellites with their early and close tracking of hurricanes, will greatly reduce the loss of life from such storms.

1976: Real-time operational forecasts and warnings using Doppler radar are evaluated by the Joint Doppler Operational Project, spawning a third Generation Weather Radar (WSR 889). The Big Thompson Canyon flood in Colorado kills 139 people.

1977: Success of weather satellites causes elimination of last U.S. weather observation ship; real-time access to satellite data by national centers advances hurricane, marine, and coastal storm forecasts.

1979–1988: Dr. Richard Hallgren appointed NOAS assistant administrator for the Weather Services.

1979: A Nested Grid Model (NGM) becomes operational; a Global Data Assimilation System (GDA) developed; AFOS Computer system is deployed connecting all Weather Service forecast offices. AFOS is the most ambitious computer network yet created, setting records for volume of data and number of entry points while supporting full range of word processing and other capabilities.

1980: Mt. St. Helens, a dormant volcano in Washington State, erupts; weather satellites spot eruption and alert FAA. "Dean of the Cooperative Weather Observers," Mr. Edward H. Stoll of Elwood, Nebraska, is honored at the nation's Capitol and meets President Jimmy Carter in the White House. Mr. Stoll had faithfully served as a cooperative observer since October 10, 1905. Various "hot weather topics" become of general public concern, such as the El Niño / Southern Oscillation (ENSO) as a factor in U.S. weather and global warming.

Did You Know?

Outside of normal seasonal variation, ENSO is one of the main sources of year-to-year variability in earth's weather and climate, with significant socioeconomic implications for many regions around the world. In normal years, trade winds push warm water and its associated heavy rainfall westward from the central Pacific toward Indonesia. During an El Niño, the winds die down and can even reverse direction, pushing the rains toward South America instead. This is why people in Indonesia and Australia typically associate El Niño with drought, while people in Peru connect El Niño with floods.

1981: Weather-related science event: World's first reusable space shuttle, Columbia, launched, completing its mission three days later.

1982: El Chicon erupts in Mexico; NOAA polar weather satellites track movement of its cloud around the earth as a possible global climate impact.

1984: The National Weather Service provides special forecast for the Olympic Games in Los Angeles. Weather-related event: First successful solo balloon crossing of the Atlantic by pilot Joe Kittinger takes 83 hours and 45 minutes.

1984, September 11–13: First official Air Transportable Mobile Unit (ATMU) dispatches to the Shasta-Trinity National Forest wildfire. The ATMU is dispatched by plane from Redding, California, while the forecaster is flying from Sacramento, California. These mobile fire units are deployed nationwide in 1987. ATMUs permit NWS forecasters to set up remote observing and forecasting offices anywhere in the world within hours of a request for on-site fire weather support.

1985: Harvard's Blue Hill Observatory celebrates 100 years of continuous monitoring of the atmosphere. President Ronald Reagan awards Dr. Helmut Landsberg the National Medal of Science, the most prestigious service award a civilian can receive.

1986: Eight-day nonstop Voyager around-the-world balloon flight completed with assistance of continuous weather support from retired, volunteer, and current Weather Service employees.

1987–1988: Major drought experienced by nation's midsection, with some of the lowest river levels in 50 years observed on the Mississippi; Dr. Hallgren retires to become president of the AMS.

1988: Weather Service operates several remote forecast operations in Yellowstone Park to assist in fighting week-long wildfire; National Hurricane Center provides continuous advisories and early forecast on movement of giant hurricane Gilbert to assist Caribbean and U.S. coastal areas with evacuation plans; Dr. Elbert W. Friday Jr. appointed NOAA assistant administrator for Weather Service.

1989: United States assists cleanup efforts in San Francisco earthquake area with mobile forecast unit; Miami Hurricane Center plays central role in limiting loss of life from gigantic storm Hugo, which causes $7 billion in damages.

Eight-year national plan for the modernization and restructuring of the National Weather Service is announced.

1990: The Cray Y-MP8 supercomputer is procured and installed at the National Meteorological Center to run higher resolution and more sophisticated numerical weather production models. The National Weather Service exercised the contract option for full-scale production with the Unisys Corporation for production of 165 Next General Radar (NEXRAD) units and over 300 display subsystems. The explosive growth of technology has led to NEXRAD, a joint project of the Departments of Commerce, Transportation, and Defense to meet their common radar needs. Continued development and planning for the Automated Surface Observing System (ASOS). Today, routine surface observations are collected manually each hour at 260 Weather Service facilities with 1,200 people giving at least part time to the task. By freeing them of manual observation, the ASOS will help provide this vital time. A joint effort of NOAA and the Federal Aviation Administration, the ASOS program will produce as many as 1,700 units—scheduled for installation at U.S. airports by the mid 1990s. Operating automatically 24 hours a day, they will alert forecasters to significant weather changes.

1991: Automated Surface Observing System contract, a key element in NOAA's modernization of the NWS, awarded to AAI Corporation of Hunt Valley, Maryland, on February 19.

1992: Twenty-two of the planned 115 modernized Weather Forecast Offices (WFO) are built or remodeled during the year, with 12 NWS radars installed. Of a programmed 1,700 ASOS units, 151 are installed and 13 commissioned. Hurricane Iniki strikes the Hawaiian island of Kauai killing seven, and Hurricane Andrew devastates Florida and Louisiana.

1993: "Year of Water"—record floods inundate the Midwest; the National Weather Service earns the U.S. Commerce Department's highest award, a gold medal, for performance during the flooding. Advanced Weather Interactive Processing System (AWIPS) contract awarded to PRC, Inc., of McLean, Virginia. AWIPS will rapidly analyze weather data and distribute it nationwide; the 100th new Doppler weather radar is installed; the Blizzard of '93 deposited enough precipitation in one weekend to drastically change the spring hydroponic outlook; an international training facility was dedicated at the National Meteorological Center. Two scientists develop a new method of processing atmospheric data needed for global forecasting, and five meteorologists from Alaska design a state-of-the-art computer network used to improve forecasting capabilities in Alaska.

1994: Dr. Elbert W. Friday Jr. is honored as Federal Executive of the Year; tornadoes plow through Southeast United States killing 40; the new GRAY C90 supercomputer is dedicated, providing faster and more accurate forecasts; NOAS and the EPA launch an experimental Ultraviolet Exposure Index.

Weather and Climate

Weather is a constant minor and, occasionally, major concern for most of us. Do you know anyone who (like Helen Keller) claims to be able to smell changes in the weather, or who says they "sense" a change in the weather brewing? You have heard people talk about the weather—have you heard anyone ask (in ordinary conversation) what the weather is like in someone else's locale? Or whether you'd heard the weather forecast for the weekend?

Ever gotten a letter that talked about how the weather in the writer's area had been or been asked how the weather in your area was? Did you know about London's Victorian-era affinity for dark-tinted wallpaper being weather related?

Because weather affects us all, physically and emotionally, we are often concerned with what changes weather is going to bring to us on a day-to-day and season-to-season basis. Whether or not you know anyone who can smell or sense weather change, as mentioned, people discuss the weather in conversation daily. We're interested in the weather conditions in other people's cities (especially bad weather), and people move to one locale or another because of that area's weather conditions. Or do they? Do they move because of the weather there—or because of the climate?

Did You Know?

Accept the fact that we will be wrong about the weather, perhaps more often than we will be right.

Let's look at climate for a moment. Have you ever asked someone how their climate is—when you really wanted to know what their weather was? We don't often confuse the two. When we talk about weather, we are generally referring to the transient changes in temperature, precipitation, and wind that affect whether we take the umbrella along or wear a heavy coat. Most people rely heavily on the local meteorologist and the daily weather forecasts: so much so that an entire (and very visible) branch of science is dedicated to the effort of trying to predict the weather—a difficult task, because of the extensive variables in any prediction.

Try to define climate and weather. Most people do not have a good feel for the exact meanings of and differences between "weather" and "climate." The two terms and their specific meanings and differences, and the elements that comprise them, are the subject of this chapter.

Meteorology: The Science of Weather

Meteorology is the science concerned with the atmosphere and its phenomena; the meteorologist observes the atmosphere's temperature, density, winds, clouds, precipitation, and other characteristics and endeavors to account for its observed structure and evaluation (weather, in part) in terms of external influence and the basic laws of physics. Meteorological phenomena affect the chemical properties of the atmosphere.

Weather is the state of the atmosphere, mainly with respect to its effect upon life and human activities; as distinguished from **climate** (long-term manifestations of weather), weather consists of the short-term (minutes to months) variations of the atmosphere. Weather is defined primarily in terms of heat (temperature), pressure, clouds, humidity, wind, and moisture—the elements of which weather is made. At high levels above the earth, where the atmosphere thins to near vacuum, weather does not exist. Weather is a near-surface phenomenon. We see this clearly, daily, as we observe the ever-changing, sometimes dramatic, and often violent weather display that travels through our environment.

In the study of environmental science, and in particular the study of air quality (especially regarding air pollution in a particular area), the determining factors are directly related to the dynamics of the atmosphere—local weather. These determining factors include strength of winds, the direction they are blowing, temperature, available sunlight (needed to trigger photochemical reactions, which produce smog), and the length of time since the last weather event (strong winds and heavy precipitation) cleared the air.

Non-destructive weather events (including strong winds and heavy precipitation) that work to clean the air we breathe are beneficial, obviously. However, few people would categorize weather events such as tornadoes, hurricanes, and typhoons as beneficial. Other weather events can have both a positive and negative effect. One such event is **El Niño**, which we discuss later in chapter 8.

Winds and Breezes

On bright clear nights, the earth cools more rapidly than on cloudy nights, because cloud cover reflects a large amount of heat back to earth, where it is reabsorbed again. The earth's air is heated primarily by contact with the warm earth. When air is warmed, it expands and becomes lighter. Air warmed by contact with earth rises and is replaced by cold air that flows in and under it. When this cold air is warmed, it too rises and is replaced by cold air. This cycle continues and generates a circulation of warm and cold air called **convection**.

At the earth's equator, the air receives much more heat than the air at the poles. This warm air at the equator is replaced by colder air flowing in from north and south. The warm, light air rises and moves poleward high above the earth. As it cools, it sinks, replacing the cool surface air that has moved toward the equator.

The circulating movement of warm and cold air (convection) and the differences in heating cause local **winds** and **breezes**. Different amounts of heat are absorbed by different land and water surfaces. Soil that is dark and freshly plowed absorbs much more than grassy fields, for example. Land warms faster than water during the day and cools faster at night. Consequently, the air above such surfaces is warmed and cooled, resulting in production of local winds.

Winds should not be confused with air currents. Wind is primarily oriented toward horizontal flow, while **air currents** are created by air moving upward and downward. Both wind and air currents have direct impact on air pollution, which is carried and dispersed by wind. An important factor in determining the locations most affected by an air pollution source is wind direction. Since air pollution is a global problem, wind direction on a global scale is important (see Figure 1.1).

Another constituent associated with earth's atmosphere is water. Water is always present in the air. It evaporates from the earth, two-thirds of which is covered by water. In the air, water exists in three states: solid, liquid, and invisible vapor.

The amount of water in the air is called **humidity**. The **relative humidity** is the ratio of the actual amount of moisture in the air to the amount needed for saturation at the same temperature. Warm air can hold more water than

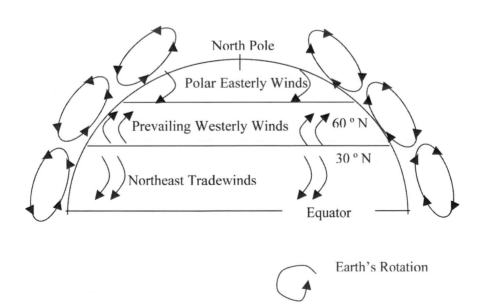

FIGURE 1.1
Global wind directions in the northern hemisphere.

cold. When air with a given amount of water vapor cools, its relative humidity increases; when the air is warmed, its relative humidity decreases.

Air Masses

An **air mass** is a vast body of air (so vast that it can have global implications) in which the conditions of temperature and moisture are much the same at all points in a horizontal direction. An air mass is affected by and takes on the temperature and moisture characteristics of the surface over which it forms, though its original characteristics tend to persist.

When two different air masses collide, a **front** is formed. A **cold front** marks the line of advance of a cold air mass from below as it displaces a warm air mass. A **warm front** marks the advance of a warm air mass as it rises up over a cold one.

Thermal Inversions and Air Pollution

We've said that during the day the sun warms the air near the earth's surface. Normally, this heated air expands and rises during the day, diluting low-lying pollutants and carrying them higher into the atmosphere. Air from surrounding high-pressure areas then moves down into the low-pressure area created when the hot air rises (see top of Figure 1.2). This continual mixing of the air helps keep pollutants from reaching dangerous levels in the air near the ground.

Sometimes, however, a layer of dense, cool air is trapped beneath a layer of less dense, warm air in a valley or urban basin. This is called a **thermal inversion** (see Figure 1.2b). In effect, a warm-air lid covers the region and prevents pollutants from escaping in upward-flowing air currents. Usually these inversions trap air pollutants at ground level for a short period of time. However, sometimes they last for several days, when a high-pressure air mass stalls over an area, trapping air pollutants at ground level where they accumulate to dangerous levels.

The best known location in the United States where thermal inversions occur almost on a daily basis is in the Los Angeles basin. The Los Angeles basin is a valley with a warm climate and light winds, surrounded by mountains located near the Pacific Coast. Los Angeles itself is a large city with a large population of people and automobiles—and L.A. possesses the ideal conditions for smog—conditions worsened by frequent thermal inversions.

Interesting Point: Earlier I stated that London's Victorian-era affinity for dark-tinted wallpaper was weather related—in fact, a clever attempt to limit the visual impact of the dirty deposits left by air pollution.

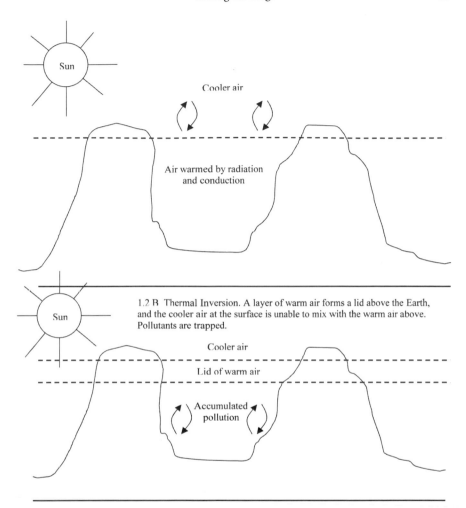

Cooler air

Air warmed by radiation
and conduction

1.2 B Thermal Inversion. A layer of warm air forms a lid above the Earth,
and the cooler air at the surface is unable to mix with the warm air above.
Pollutants are trapped.

Cooler air

Lid of warm air

Accumulated
pollution

FIGURE 1.2
Normal conditions (top). Air at the earth's surface is heated by the sun and rises to mix
with the cooler air above it. Thermal inversion is bottom, B. A layer of warm air forms
a lid above the earth, and the cooler air at the surface is unable to mix with the warm
air above. Pollutants are trapped.

References and Recommended Reading

Ackerman, D. 1990. *A Natural History of the Senses.* New York: Random House.
Anthes, R. A., Cahir, J. J., Frasier, A. B., & Panofsky, H. A. 1984. *The Atmosphere,* 3rd
    ed. Columbus, OH: Charles E. Merrill Publishing Co.
Battan, L. J. 1983. *Weather in Your Life.* New York: W. H. Freeman.
Budyko, M. I. 1982. *The Earth's Climate.* New York: Academic Press.

Gates, D. M. 1962. *Energy Exchange in the Biosphere.* New York: Harper & Row Monographs.

Ingersoll, A. P. 1983. "The Atmosphere." *Scientific American* 249(3): 162–174.

Kondratyev, K. Y. 1969. *Radiation in the Atmosphere.* New York: Academic Press.

Moran, J. M., & Morgan, M. D. 1994. *Essentials of Weather.* Upper Saddle River, NJ: Prentice-Hall.

National Research Council. 1975. *Understanding Climatic Change: A Program for Action.* Washington, DC: National Academy of Sciences.

National Research Council. 1982. *Solar Variability, Weather, and Climate.* Washington, DC: National Academy Press.

NOAA. 1976. *U.S. Standard Atmosphere.* NOAA S/T 76-1562.

NOAA. 2007. *Weather Maps.* Accessed May 22, 2011 at http://www.nssl.noaa.gov/edu/lessons/lesson_map.html.

NOAA. 2010. *Evolution of the National Weather Service.* Public Affairs Office Publication.

II

METEOROLOGICAL BASICS

From my wings are shaken the dews that waken
The sweet buds every one,
When rocked to rest on their mother's breast,
As she dances about the Sun.

—Percy Bysshe Shelley, *The Cloud*, 1820

2

Weather Elements

Whether the weather be fine
Or whether the weather be not,
Whether the weather be hot,
We'll weather the weather
Whatever the weather,
Whether we like it or not.

—Anonymous

(MUCH OF THE INFORMATION in this introductory section is from NASA's *How Atmospheric Pressure Affects the Weather*. Accessed June 27, 2011, at http://kids.earth.nasa.gov/archive/air_pressure/barometer.html.) What affects the weather? What causes rain, snow, hail, sunny days, cloudy days, windy days, and just plain old nasty weather days? How do we know when the weather will change; when will it get better or get worse? In answering these questions we begin by stating that weather is affected by certain weather elements. Well, this raises another question: What is a weather element? Simply, there are nine basic weather elements and they are defined as elements that can be measured in the atmosphere: air pressure, precipitation, wind, air temperature, relative humidity, solar energy, evaporation, pollution, and cloud heights. Okay, how about the other questions: What causes rain, snow, hail, sunny days, cloudy days, windy days, and just plain old nasty weather days? How do we know when the weather will change or when it will get better or worse? Before we can answer these questions and many others we must first understand all nine weather elements and how they work to affect weather in general.

Air Pressure

Air pressure is the amount of force the atmosphere exerts on a surface. More specifically we can say that air pressure is the weight of the air over and around our bodies. It affects ensuing winds and the probability of rain. The most fundamental thing to know about air pressure is that heavier gases weigh more than lighter gases. So what, you ask? Well, different chemical elements, as you know, have different atomic weights. Those that form gases (like nitrogen, oxygen, etc.) often combine two atoms at a time to form a gaseous molecule—like N_2 (two nitrogens) or O_2 (two oxygens).

The atomic weight of nitrogen (N) is 14; for oxygen (O) it's 16. The molecules N_2 and O_2 have molecular weights of 28 and 32, respectively. Obviously, a gallon of oxygen weighs more than a gallon of nitrogen. Moreover, it turns out that at room temperature and normal (sea level) atmospheric pressure, 28 grams of nitrogen occupies a volume of 22.4 liters and 32 grams of oxygen occupies the same volume.

The point is under STP (standard temperature and pressure) the weight of 22.4 liters of a gas in grams equals the molecular weight of the gas. Consider the following example:

Example 2.1

Problem

We know that air is about 80% nitrogen and 20% oxygen. How much does a liter of air weigh?

Solution

If 22.4 liters of nitrogen weighs 28 grams; 0.8 liters weighs $(0.8/22.4) \times 28$ grams = 1 gram (almost exactly) and 0.2 liters of oxygen weighs $(0.2/22.4) \times 32 = 0.286$ grams—thus: A liter of air weighs about 1.286 grams.

Let's look at another example.

Example 2.2

Problem

How much does gaseous water weigh? Not liquid water—steam or vapor.

Solution

The chemical formula for water is H_2O: one oxygen atom (atomic weight 16) and two hydrogen atoms (atomic weight 1). The total weight of the molecule is 18. So 22.4 liters (of the gas) weighs 18 grams. One liter weighs 18/22.4 grams or 0.8 grams. Again, we know that the weight of air is 1.286 grams per liter, but if we substitute water for some of the air, the mixture becomes light. So, if there's water (humidity) in the air, the air mixture becomes lighter.

Recall that we stated that a weather element, such as air pressure, is something that we can measure in the atmosphere. Because air at a higher temperature is less dense than the cooler air over another air, it rises. This movement, in turn, generates difference in pressure between different areas. To measure the weather element pressure we use a barometer.

A barometer, based on the principle developed by Evangelista Torricelli in 1643, which measures the weight of atmosphere above it, will show a lower pressure under a rising column of warm air, which displaces the cooler, relatively denser air above it. On the other hand, a descending volume of cool air will cause higher pressure below it. In regard to the barometer, air pressure is measured by observing the height of the column of mercury in the tube. At sea level, air pressure will push on the mercury (in a liquid mercury barometer) at the open end and support a column of mercury about 30 inches high. As atmospheric pressure increases, the mercury is forced from the reservoir by the increasing air pressure and the column of mercury rises; when the atmospheric pressure decreases, the mercury flows back into the reservoir and the column of mercury is lowered.

Did You Know?

If water were used instead of mercury in a barometer, a glass tube over 30 feet high would need to be used.

In regard to actual conditions, the pressure difference at the surface causes a wind to blow from the region of higher pressure toward that of lower pressure. It is similar to what happens when an inflated balloon is punctured at one end. The high-pressure air inside rushes out into the surrounding area of lower pressure. The greater the pressure difference, the stronger the wind. This principle operates on a global scale.

Key Points

- Air pressure is measured by a liquid mercury barometer or an aneroid barometer.

- Pressure $= \dfrac{\text{force}}{\text{area}} = \dfrac{\text{lb}}{\text{in}^2} = \dfrac{\text{dynes}}{\text{cm}^2}$ or millibars (mb)
- Standard pressure $= 14.7 \text{ lb/in}^2 = 1{,}013{,}250 \text{ dynes/cm}^2 = 1013.25 \text{ mb}$
- 1 in. Hg $= 0.49116/\text{lb.in}^2 = 33{,}863.9 \text{ dynes/cm}^2 = 33.8639 \text{ mb}$

Did You Know?

The aneroid barometer is not as precise as a mercury barometer but is cheaper, lighter, and less poisonous.

Before hurricanes could be spotted by satellites from space, people would keep a wary eye on their barometers during hurricane season. If the air pressure dropped, that was usually a good time to board up windows and head further inland.

Warm air near the equator rises and moves toward the poles. In turn, the cooler polar air is drawn beneath, toward the equator. Keep in mind that air is always trying to equalize itself over the world, thus, as mentioned, is trying to leave an area of high pressure to get to a low-pressure area. The earth's rotation diverts these currents into prevailing easterly and westerly winds at different latitudes and altitudes.

These global movements of air masses in different directions generate turbulent motions where the main flow streams brush against each other. Topography also causes additional effects—the irregular outlines of continents and the complex patterns of mountains and plains, deserts and forests.

When the weather changes, air pressure changes and causes the barometer to change. If the barometer is rising, the air is cooling and expanding and there is a tendency for fair weather. If it is dropping, the air is being compressed and is getting warmer, so stormy weather could result.

Precipitation

One of the simplest weather elements to measure is precipitation. Generally, this is accomplished using some type of rain gauge. Rain gauges come in a variety of shapes and sizes, but all are used to accurately measure the amount of fallen precipitation. The simplest rain gauge is a can with vertical walls and a ruler. The typical home rain gauge consists of a plastic tube or wedge with measurement increments on its side. One just reads the water level on the side of the rain gauge.

According to the National Weather Service (2011), the most common rain gauge found at weather offices is the standard 8-inch rain gauge that has been used for over 100 years. It consists of a large cylinder with a funnel and a smaller measuring tube inside of it (as shown in Figure 2.1 and Example 2.3). The dimensions of this instrument are very specific so water that collects in the measuring tube has exactly one-tenth the cross sectional area of the top of the funnel. The reason for the smaller measuring tube is so that more precise rainfall would actually fill an inch of the measuring tube. A special measuring stick inserted into the measuring tube takes into account the vertical scale exaggeration.

Example 2.3

An alternative to the standard rain gauge for measuring rainfall is the tipping bucket. It uses gravity. Two specially designed buckets tip when the weight of 0.1 inches of rain falls into them. When one bucket tips, the other bucket quickly moves into place to catch the rain. Each time a bucket tips, an

8" in. Diameter

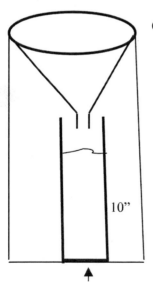

Circumference of a circle $= 2\pi R$

Circumference of example $= 2\pi 4$"

$= 3.14 \times 8$"

$C = 2.53 = 2.53/10$

(by dividing by ten we can measure directly)

10"

2.53" = 1/10 the length of the funnel opening with an 8" diameter.

FIGURE 2.1
For Example 2.3.

electronic signal is sent to a recorder. To calculate the rainfall for a certain time period, simply multiply the number of marks on the recorder by 0.01 inches. The tipping bucket rain gauge is especially good at measuring drizzle and very light rainfall events; unfortunately it has a difficult time keeping up with heavy rain from a thunderstorm.

The weighing rain gauge is another type of rain gauge. It is composed of a container sitting upon a scale. The scale is adjusted for the container and measures the weight of the collected rain water. The measurement is traced out on a rain chart for a permanent rain record.

Did You Know?

Snow gauges are similar to rain gauges but with a few minor variations. First, they are placed higher than ground level, usually in trees with the can wide open. The weight of the snow and the water is measured, then a propane heater melts the snow, and a radio transmitter relays the information to a National Weather Service receiver. Standard conversion rules for converting snow level to actual water level in inches are as follows:

- Wet snow = 10" or less of snow to 1" of water
- Normal snow = 12" of snow to 1" of water
- Dry snow = 15" (up to 30") of snow to 1" water

Wind

Wind is a weather element with two measurements that concern meteorologists: wind direction and wind speed. In regard to wind direction, we are all familiar with the weather vane (see Figure 2.2), which is designed to turn itself into an oncoming wind; it tells us the direction the wind is coming *from*. Keep in mind that all wind direction information tells us where the wind is coming from, not where it is going.

Wind speed, or wind velocity, is a fundamental atmospheric rate and is measured by an anemometer (see Figure 2.3); it can also be classified using the older Beaufort scale, which is based on people's observations of specifically defined wind effects.

As shown in Figure 2.3, the rotating cup anemometer usually consists of three hemispherical or cone-shaped cups mounted symmetrically about a vertical axis of rotation. The rate of rotation of the cups is essentially linear over the normal range of measurements, with the linear wind speed being about 2 to 3 times the linear speed of a point on the center of a cup, depending on the construction of the cup assembly.

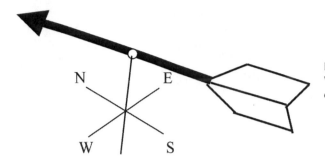

FIGURE 2.2
Weather vane—weighted
end faces into the wind.

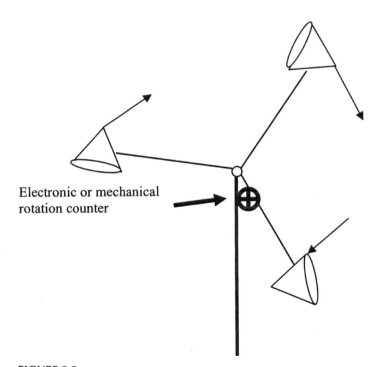

Electronic or mechanical
rotation counter

FIGURE 2.3
Cup anemometer. The mounted electronic or mechanical rotation-
counter counts or signals rotations as wind turns the cups. By
counting the signals and measuring the radius of the arm, the
circumference of the circle the distance the wind traveled is found.
Thus, mathematically we can state wind speed as a result of distance/
time = MPH.

Another type of anemometer commonly used today is the vane-oriented propeller anemometer. It usually consists of a two-, three-, or four-bladed propeller that rotates on a horizontal pivoted shaft that is turned into the wind by a vane. There are several propeller anemometers that employ light-weight molded plastic or polystyrene foam for the propeller blades to achieve low stating threshold speeds. Some propeller anemometers are not associated with a moving vane. Rather, two orthogonal fixed-mount propellers are used to determine the vector components (i.e., speed and direction) of the horizontal wind. A third propeller with a fixed mount rotating about a vertical axis may be used to determine the vertical component of the wind if desired.

Air Temperature

Temperature, another weather element, is a measure of how "hot" something is or how much thermal energy it contains. Specifically, it is the kinetic energy of the atmosphere and is measured with a thermometer. Temperature is a fundamental measurement in air science, especially in most pollution work. The temperature of a stack gas plume, for example, determines its buoyancy and how far the plume of effluent will rise before attaining the temperature of its surroundings. This, in turn, determines how much it will be diluted before traces of the pollutant reach ground level.

Temperature is measured on several scales: For example, the *centigrade* (*Celsius*) and *Fahrenheit* scales are both measured from a reference point— the freezing point of water—which is taken as 0°C or 32°F. The boiling point of water is taken as 100°C or 212°F. For thermodynamic devices, it is usual to work in terms of absolute or thermodynamic temperature where the reference point is absolute zero, which is the lowest possible temperature attainable. For absolute temperature measurement, the thermodynamic unit or *Kelvin* (K) scale is used, the Kelvin scale uses centigrade divisions for which zero is the lowest attainable measurement. A unit of temperature on this scale is equal to a Celsius degree but is not called a degree; it is called a Kelvin and is designated as K, not °K. The value of absolute zero on the Kelvin scale is −273.15°C, so that the Kelvin temperature is always a number 273 (rounded) higher that the Celsius temperature. Thus water boils at 373 K and freezes at 273 K.

It is easy to convert from the Celsius scale to the Kelvin scale. Simply add 273 to the Celsius temperature and you have the Kelvin temperature. Mathematically,

$$K = C + 273$$

where K = temperature on the Kelvin scale and C = temperature on the Celsius scale. Converting from Fahrenheit to Celsius or vice versa is not so easy. The equations used are

$$C = 5/9(F - 32) \text{ and } F = 9/5C + 32$$

where C = temperature on the Celsius scale and F = temperature on the Fahrenheit scale. As examples, 15°C = 59°F and 68°F = 20°C. F or C of course, can be negative numbers.

Thermometers

Over the years many different devices (thermometers) have been developed to measure temperature. In late 1593, well before making his other inventions and discoveries, Galileo invented the thermometer but no standardization of measurement. In 1714 Fahrenheit devised a temperature scale. To make any scale, two ends are selected as standards and then the in-between readings are calibrated. Fahrenheit chose the lowest or coldest temperature that could be reached in his day (that is, the freezing point of sal ammoniac), as 0°, and then used the temperature of his own body to be 100°, found ice to be 32° and boiling water 212°. Consequently, there are 180° between the boiling point and freezing point of water making the degrees, relative to centigrade degrees, small. In 1742, Anders Celsius (1701–1744) put forth the centigrade temperature scale. He used water as the standard of temperature: The freezing point of distilled water at sea level equals 0° and the boiling point equals 100°. Consequently, there are 100° between the boiling point and freezing point of water in the Celsius scale, making the centigrade degree almost twice as large as the Fahrenheit degree. Finally, Lord Kelvin (1824–1907) established the absolute scale, setting absolute zero, where, theoretically, all molecular motion is lost. Every time the temperature is lowered 1 Kelvin, the absolute (molecular) pressure is lowered by 1/273. As mentioned, the Kelvin is the same size as the centigrade degree, thus: 0° Celsius = 273 Kelvin or K = C° + 273.

Did You Know?

In liquid glass thermometers alcohol is used to read lower temperatures because the freezing limit of alcohol is −170°F and that of mercury is −34°F. However, mercury is more reliable because it isn't so pressure sensitive and won't evaporate.

In meteorology, the temperature changes we are concerned with are, of course, derived from the sun. If left in the sun, thermometers measure absorption qualities of the materials of which they are made. Color, roughness, and material can affect the temperature indicated if the thermometer is left in the sun. In temperature recording activities for official purposes, all temperatures are recorded in the shade to prevent recording of the conductivity of the glass or metal of the thermometer and not the heat of the air mass.

The sun is the weather generator. Thus, it should not be surprising that inventors have been trying to come up with some type of sunshine recorder—to track sunlight during a daylight period. In 1853, John Campbell did just that (later modified by Stokes in 1879); they invented the Campbell-Stokes recorder. It is a sunshine recorder—probably the most common—that is still in use today. Their recorder consisted of a crystal ball with a north–south axis lined up with that of the earth and placed in an open, sunlit location. On the back of the device is a strip of paper placed in such a manner that as the sun moves across the sky, it burns a trace on the paper (if clouds, no burn).

Did You Know?

Meteorological instruments placed in outdoor field locations, such as thermometers, must be enclosed to shield the instrument(s) against precipitation and direct heat radiation from outside sources, while still allowing air to circulate freely around them. The enclosures commonly used are the Stevenson screen or a cotton region shelter. These shelters are parts of standard weather stations that provide a standardized environment in which to measure temperature, humidity, dew point, and atmospheric pressure.

Relative Humidity

The weather element relative humidity is a dimensionless ratio, expressed in percent, of the amount of atmospheric moisture (water vapor, a true gas) present relative to the amount that would be present if the air were saturated. Because the latter amount is dependent on temperature, relative humidity is a function of both moisture content and temperature. As such, relative humidity by itself does not directly indicate the actual amount of atmospheric moisture present; moreover, it is difficult to measure accurately.

In measuring humidity, the psychrometer and sling psychrometer are used to measure the coolness of a wet bulb. The principle behind this is that when water evaporates, it absorbs energy from its surroundings in order to change phase. This has a cooling effect on the immediate environment that will increase in intensity proportionate to the availability of the surrounding air to hold water molecules. A simple example is to lick your finger and hold it in the air or move it about in the air. It takes a minute or two for the psychrometer to measure the change. To use the psychrometer, spin it or use a fan and cover the wet bulb with several wrappings of clean muslin and wet with distilled water. Read the wet bulb first, subtract from the dry bulb, and then check variance humidity or dew point tables available from NOAA, NWS, and other agency or organization sources.

Today, modern electronic devices have replaced many of the older instruments that were used for years to measure humidity. These modern electronic devices use temperature and condensation, changes in electrical resistance, and changes in electrical capacitance to measure humidity changes.

Solar Energy

> All living creatures and all plants derive their life from the sun. If it were not for the sun, there would be darkness and nothing could grow—the earth would be without life.
>
> —Okute, Teton Sioux

The sun is the driving force behind weather. Without the distribution and reradiation to space of solar energy, we would experience no weather (as we know it) on earth. The sun is the source of most of the earth's heat. Of the gigantic amount of solar energy generated by the sun, only a small portion bombards the earth. Most of the sun's solar energy is lost in space. A little over 40% of the sun's radiation reaching earth hits the surface and is changed to heat. The rest stays in the atmosphere or is reflected back into space.

Like a greenhouse, the earth's atmosphere admits most of the solar radiation. When solar radiation is absorbed by the earth's surface, it is reradiated as heat waves, most of which are trapped by carbon dioxide and water vapor in the atmosphere, which work to keep the earth warm in the same way a greenhouse traps heat.

The atmosphere plays an important role in regulating the earth's heating supply. The atmosphere protects the earth from too much solar radiation during the day and prevents most of the heat from escaping at night. Without the filtering and insulating properties of the atmosphere, the earth would experience severe temperatures similar to other planets.

On bright clear nights the earth cools more rapidly than on cloudy nights because cloud cover reflects a large amount of heat back to earth, where it is reabsorbed.

The earth's air is heated primarily by contact with the warm earth. When air is warmed, it expands and becomes lighter. Air warmed by contact with earth rises and is replaced by cold air, which flows in and under it. When this cold air is warmed, it too rises and is replaced by cold air. This cycle continues and generates a circulation of warm and cold air, which is called *convection*.

Evaporation

Evaporation is another weather element commonly used to measure weather. To measure evaporation a 48-inch diameter and 10-inch deep pan is filled with water, and then the rate of evaporation of the water is measured. One significant measurement obtained by using the evaporation pan method is the determination of whether there is a net loss of water level from evaporation over rainfall. Obviously, if the loss of water exceeds the input of water from rainfall, this indicates the need to import water.

Pollution

Air pollution affects weather directly and indirectly. For example, acid rain, global warming, and weather inversion pollution problems are all weather related, caused, at least to some degree, by air pollution. We are concerned about air pollution for a number of reasons and, therefore, it is no surprise that in order to gauge local air quality, we incorporate the use of many environmental sampling devices to collect and test representative samples.

Cloud Heights

The cloud height is the distance between the cloud base and the cloud top. Cloud height is often related to the intensity of precipitation generated by a cloud: Deep clouds tend to produce more intense rainfall. For instance, cumulonimbus clouds can develop vertically through a substantial part of the troposphere and often result in a thunderstorm with lightning and heavy showers. By contrast, very thin clouds (such as cirrus clouds) do not generate any precipitation at the surface (Huschke, 1959). To measure cloud heights, a balloon with a specific quantity of helium and a known weight is launched and timed with a stop watch.

Gas Laws

To understand the physics of air, it is imperative to have an understanding of various physical laws that govern the behavior of pressurized gases. One of the more well-known physical laws is *Pascal's law*. Pascal's law states that a confined gas (fluid) transmits externally applied pressure uniformly in all directions, without change in magnitude. This parameter can be seen in a container that is flexible; it will assume a spherical (balloon) shape. However, you probably have noticed that most compressed-gas tanks are cylindrical in shape (which allows use of thinner sheets of steel without sacrificing safety) with spherical ends to contain the pressure more effectively.

Boyle's Law

Though gases are compressible, note that, for a given mass flow rate, the actual volume of gas passing through the system is not constant within the system due to changes in pressure. This physical property (the basic relationship between the pressure of a gas and its volume) is described by *Boyle's law* (named for its discoverer, Irish physicist and chemist Robert Boyle in 1662), which states, "The absolute pressure of a confined quantity of gas varies inversely with its volume, if its temperature does not change." For example, if the pressure of a gas doubles, its volume will be reduced by a half; that is, *as pressure goes up, volume goes down*. The converse is also true. This means, for example, that if 12 ft^3 of air at 14.7 psia is compressed to 1 ft^3, air pressure will rise to 176.4 psia, as long as air temperature remains the same. This relationship can be calculated as follows:

$$P_1 \times V_1 = P_2 \times V_2 \qquad (2.1)$$

where P_1 = original pressure (units for pressure must be absolute), P_2 = new pressure (units for pressure must be absolute), V_1 = original gas volume at pressure P_1, and V_2 = new gas volume at pressure P_2. This equation can also be written as

$$\frac{P_2}{P_1} = \frac{V_1}{V_2} \text{ or } \frac{P_1}{P_2} = \frac{V_2}{V_1} \qquad (2.2)$$

To allow for the effects of atmospheric pressure, always remember to convert from gage pressure *before* solving the problem, then convert back to gage pressure *after* solving it:

$$\text{Psia} = \text{psig} + 14.7 \text{ psi, and psig} = \text{psia} - 14.7 \text{ psi}$$

Note that in a pressurized gas system where gas is caused to move through the system by the fact that gases will flow from an area of high pressure to that of low pressure, we will always have a greater actual volume of gas at the end of the system than at the beginning (assuming the temperature remains constant).

Let's take a look at a typical gas problem using Boyle's law:

Example 2.4

Problem

What is the gage pressure of 12 ft³ of air at 25 psig when compressed to 8 ft³?

Solution

$$25 \text{ psig} + 14.7 \text{ psi} = 39.7 \text{ psia}$$

$$P_2 = P_1 \times \frac{V_1}{V_2} = 39.7 \times \frac{12}{8} = 59.6 \text{ psia}$$

$$\text{Psig} = \text{psia} - 14.7 \text{ psi} = (59.6 \text{ psia}) - (14.7 \text{ psi}) = 44.9 \text{ psig}$$

The gage pressure is 44.9 psig (remember that the pressures should always be calculated on the basis of absolute pressures instead of gage pressures).

Charles's Law

Another physical law dealing with temperature is *Charles's law* (discovered by French physicist Jacques Charles in 1787). It states that "the volume of a given mass of gas at constant pressure is directly proportional to its absolute temperature." (The temperature in Kelvin [273 + °C] or Rankine [absolute zero = −460°F, or 0°R].) This is calculated by using the following equation:

$$P_2 = P_1 \times \frac{T_2}{T_1} \tag{2.3}$$

Charles's law also states, "If the pressure of a confined quantity of gas remains the same, the change in the volume (V) of the gas varies directly with a change in the temperature of the gas," as given in the following equation:

$$V_2 = V_1 \times \frac{T_2}{T_1}$$ (2.4)

Ideal Gas Law

The *ideal gas law* combines Boyle's and Charles's laws because air cannot be compressed without its temperature changing. The ideal gas law is expressed by this equation:

$$\frac{P_1 \times V_1}{T_1} = \frac{P_2 \times V_2}{T_2}$$ (2.5)

Note that the ideal gas law is still used as a design equation even though the equation shows that the pressure, volume, and temperature of the second state of a gas are equal to the pressure, volume, and temperature of the first state. In actual practice, however, other factors such as humidity, heat of friction, and efficiency losses all affect the gas. Also, this equation uses absolute pressure (psia) and absolute temperatures (°R) in its calculations.

In air science practice, the importance of the ideal gas law cannot be overstated. It is one of the fundamental principles used in calculations involving gas flow in air pollution–related work. This law is used to calculate actual gas flow rates based on the quantity of gas present at standard pressures and temperatures. It is also used to determine the total quantity of that contaminant in a gas that can participate in a chemical reaction.

- Number of moles of gas
- Absolute temperature
- Absolute pressure

In practical applications, practitioners generally use the following standard ideal gas law equation:

$$V = \frac{nRT}{P} \text{ or } PV = nRT$$ (2.6)

where V = volume, n = number of moles, R = universal gas constant, T = absolute temperature, and P = absolute pressure.

Did You Know?

A mole is a measure of the number of molecules present. The value of R depends on the units used for the other parameters.

Example 2.5

Problem

What is the volume of 1 pound mole (denoted "lb mole") of combustion gas at an absolute pressure of 14.7 psia and a temperature of 68°F? (These are EPA-defined standard conditions.)

Solution

$$V = \frac{nRT}{P}$$

1. Convert the temperature from relative to absolute scale (from °F to °R).

$$T_{Absolute} = 68°F + 460 = 528°R$$

2. Calculate the gas volume.

$$1 \text{ lb mole} \times \frac{10.73 \text{ (psia) (ft}^3)}{\text{(lb mole)}} \times 528°R$$

$$V = \frac{(°R)}{14.7 \text{ psia} = 385.4 \text{ ft}^3}$$

Notes:

- The pressure exerted by a mass of air is directly proportional to the density of the air and the temperature. Density is related to volume and mass. If mass is constant, when volume increases, density decrease; when volume decrease, density increases.

- As density in a closed volume increases, pressure increases.
- As temperature in a closed volume increases, pressure increases.
- If pressure is constant, when density increases, temperatures decrease; and if density decreases, temperature increases.

Flow Rate

Gas flow rate is a measure of the volume of gas that passes a point in an industrial system during a given period of time. The ideal gas law tells us that this gas flow rate varies depending on the temperature and pressure of the gas stream and the number of moles of gas moving per unit of time.

When gas flow rates are expressed at actual conditions of temperature and pressure, the actual gas flow rate is being used. Gas flow rates can also be expressed at standard conditions of temperature and pressure; this is referred to as the standard gas flow rate.

Gas Conversions

Gases of interest in air pollution control are usually mixtures of several different compounds. For example, air is composed of three major constituents: nitrogen (N_2) at approximately 78.1%, oxygen (O_2) at approximately 20.9%, and argon at 0.9%. Many flue gas streams generated by industrial processes consist of the following major constituents: (1) nitrogen, (2) oxygen, (3) argon, (4) carbon dioxide (CO_2), and (5) water vapor (H_2O). Both air and industrial gas streams also contain minor constituents, including air pollutants, present at concentrations that are relatively low compared to these major constituents.

Did You Know?

Argon is usually not listed in most industrial gas analyses because it is chemically inert and difficult to measure. The argon concentration is often combined with the nitrogen concentration to yield a value of 79.0%.

There is a need for ways to express both the concentrations of the major constituents of the gas stream and the concentrations of the pollutants present as minor constituents at relatively low concentrations. There are a variety of ways to express gas phase concentrations, which can easily be converted from one type of unit to another. They include the following:

Major Constituents

Volume percent is one of the most common formats used to express the concentrations of major gas stream constituents such as oxygen, nitrogen, carbon dioxide, and water vapor. The format is very common partially because the gas stream analysis techniques used in EPA emissions testing methods provide data directly in a volume percent format.

Partial pressure concentrations can also be expressed in terms of partial pressures. This expression refers to the part of the total pressure exerted by one of the constituent gases.

Gases composed of different chemical compounds such as molecular nitrogen and oxygen behave physically the same as gases composed of a single compound. At any given temperature, one mole of a gas exerts the same pressure as one mole of any other type of gas. All of the molecules move at a rate that is dependent on the absolute temperature, and they exert pressure. The total pressure is the sum of the pressures of each of the components. The equations below are often called Dalton's law of partial pressures.

$$P_{Total} = P_i + P_{ii} + P_{iii} \cdots P_n \tag{2.7}$$

$$P_{Total} = \sum_{i=1}^{n} P_i \tag{2.8}$$

$$\text{Partial pressure (gas)} = [\frac{\text{volume \% (gas)}}{100 \text{ \%}}] \times P_{Total} \tag{2.9}$$

Because the partial pressure value is related to the total pressure, concentration data expressed as partial pressure are not the same at actual and standard conditions. The partial pressure values are also different in American engineering units and cgs units.

Both Major and Minor Constituents

Mole fraction is simply an expression of the number of moles of a compound divided by the total number of moles of all the compounds present in the gas.

Minor Constituents

- Parts per million (ppm)
- Milligrams per cubic meter (mg/m³)
- Micrograms per cubic meter (µg/m³)
- Nanograms per cubic meter (ng/m³)

All of the concentration units above can be expressed in a dry format as well as corrected to a standard oxygen concentration. These corrections are necessary because moisture and oxygen concentrations can vary greatly in gas streams, causing variations in pollutant concentrations.

Gas Density

Gas density is important primarily because it affects the flow characteristics of the moving gas streams. Gas density affects the velocities of gas through ductwork and air pollution control equipment. It determines the ability to move the gas stream using a fan. Gas density affects the velocities of gases emitted from the stack and thereby influences the dispersion of the pollutants remaining in the stack gases. It affects the ability of particles to move through gases. It also affects emission testing. Gas density data are needed in many of the calculations involved in air pollution control equipment evaluation, emissions testing, and other air pollution control–related studies.

As discussed earlier, the volume of a gas increases as the temperature increases due to the motion of the gas molecules. As the volume occupied by the gas increases, its density decreases. Density is the mass per unit volume as indicated in the following equation.

$$P_{(T=i, P=j)} = \frac{m}{V_{(T=I, P=j)}} \qquad (2.10)$$

where $P_{(T=I, P=j)}$ = density at T = I, P = j, m = mass of a substance, $V_{(T=I, P=j)}$ = volume at T = I, P = j, T = absolute temperature, and P = absolute pressure.

Did You Know?

Gas density is expressed as the mass per unit of volume of gas. The gas volume is always expressed at actual conditions. The gas volume is not corrected for temperature, pressure, moisture, or oxygen levels.

Heat Capacity and Enthalpy

The *heat capacity* of a gas is the amount of heat required to change the temperature of a unit-mass of gas one temperature degree. *Enthalpy* represents the total quantity of internal energy, such as heat, measured for a specific quantity of material at a given temperature. Enthalpy data are often represented in units of energy (e.g., Btu, kcal, joule, etc.). The enthalpy content change is often expressed in Btu/unit mass (Btu/lb_m) or Btu/unit time (Btu/SCF). The change in enthalpy of the total quantity of material present in a system is expressed in units of Btu/unit time (BTU/min). The symbols, H and ΔH, denote enthalpy and the change in enthalpy, respectively.

Heat and Energy in the Atmosphere

In addition to the importance of heat on a particular airstream, it is important to point out that heat also has an impact on earth's atmosphere and thus on atmospheric science. The sun's energy is the prime source of earth's climatic system. From the sun, energy is reflected, scattered, absorbed, and reradiated within the system but without uniform distribution. Some areas receive more energy than they lose; in some areas the reverse occurs. If this situation were able to continue for long, the areas with an energy surplus would get hotter—too hot; and those with a deficit would get colder—too cold. This does not happen because the temperature differences produced help to drive the wind and ocean currents of the world. They carry heat with them, either in the sensible or latent forms, and help to counteract the radiation imbalance. Winds from the tropics are, therefore, normally warm, carrying excess heat with them. Polar winds are blowing from areas with a deficit of heat and so are cold. Acting together, these energy transfer mechanisms help to produce the present climates on earth.

Adiabatic Lapse Rate

The atmosphere is restless, always in motion either horizontally or vertically or both. As air rises, pressure on it decreases and in response it expands. The act of expansion to encompass its new and larger dimensions requires an expenditure of energy; since temperature is a measure of internal energy, this use of energy makes its temperature drop—this is an important point—an important process in physics (especially in air physics).

This phenomenon is known as the *adiabatic lapse rate*. Simply, **adiabatic** refers to a process that occurs with or without loss of heat, especially the expansion or contraction of a gas in which a change takes place in the pressure or volume, although no heat is allowed to enter or leave.

Lapse rate refers to the rate at which air temperature decreases with height. The normal lapse rate in stationary air is on the order of 3.5°F/1,000 ft (6.5°C/km). This value may vary with latitude and changing atmospheric conditions (e.g., seasonal changes). A parcel of air that is not immediately next to the earth's surface is sufficiently well insulated by its surroundings that either expansion or compression of the parcel may be assumed to be adiabatic. The air temperature may be calculated for any height by the general formula

$$T = T_0 - Rh \tag{2.11}$$

where T = temperature of the air, h = height of the air, T_0 = temperature of the air at the level from which the height is measured, and R = lapse rate.

Example 2.6

If the air temperature of stationary air (R = 3.5°F/1,000 ft) at the earth's surface is 70°F, then at 5,000 feet, the stationary air temperature would be

$$T = T_0 - Rh$$
$$= 70°F - (3.5°F/1,000 \text{ ft}) (5,000 \text{ ft})$$
$$= 70°F - 17.5°F = 52.5°F$$

In this case, the formula simply says that for every 1,000 feet of altitude (height), 3.5° is subtracted from the initial air temperature.

Adiabatic lapse rates have an important relationship with atmospheric stability and will be discussed in greater detail later in the text.

Viscosity

All fluids (gases included) resist flow. *Absolute viscosity* is a measure of this resistance to flow. The absolute viscosity of a gas for given conditions may be calculated from the following formula:

$$\mu = 51.12 + 0.372(T) + 1.05 \times 10^{-4}(T)^2 \tag{2.12}$$
$$+ 53.147 \ (\% \ O_2/100\%) - 74.143 \ (\% \ H_2O/100\%)$$

where μ = absolute viscosity of gas at the prevailing conditions, micropoise; T = gas absolute temperature, °K; % O_2 = oxygen concentration, % by volume; and % H_2O = water vapor concentration, % by volume.

As this equation indicates, the viscosity of a gas increases as the temperature increases. It's harder to push something (e.g., particles) through a hot gas stream than a cooler one due to increased molecular activity as temperature rises, which results in increased momentum transfer between the molecules. For liquids, the opposite relationship between viscosity and temperature holds. The viscosity of a liquid decreases as temperature increases. It's harder to push something through a cold liquid than a hot one because in liquids, hydrogen bonding increases with colder temperatures.

Did You Know?

Gas viscosity actually increases very slightly with pressure, but this variation is very small in most air pollution–related engineering calculations.

The absolute viscosity and density of a gas are occasionally combined into a single parameter since both of these parameters are found in many common equations describing gas flow characteristics. The combined parameter is termed the kinematic viscosity. It is defined in the following equation.

$$u = \mu/p \qquad (2.13)$$

where u = kinematic viscosity, m^2/sec, μ = absolute viscosity, Pa•sec, and p = gas density, gm/cm^3. The kinematic viscosity can be used in equations describing particle motion through gas streams. The expression for kinematic viscosity is used to simplify these calculations.

References and Recommended Reading

EPA. 2007. Basic Concepts in Environmental Sciences: Modules 1 and 2. Accessed December 30, 2007, at www.epa.gov/apti/bces/home/index.htm.
Heinlein, R. A. 1973. *Time Enough for Love.* New York: G.P. Putnum's Sons.
Hesketh, H. E. 1991. *Air Pollution Control: Traditional and Hazardous Pollutants.* Lancaster, PA: Technomic Publishing Company.
Huschke, R. E. 1959. *Glossary of Meteorology*, 2nd ed. Boston: American Meteorological Society.
NASA. 2007. *Pascal's Principle and Hydraulics.* Accessed December 29, 2007, at www.grc.nasa.gov/WWW/k-12/WindTunnel/Activities/Pascals_principle.html.
Spellman, F. R., & Whiting, N. 2006. *Environmental Science & Technology: Concepts and Applications.* Rockville, MD: Government Institute.

3

The Atmosphere

This most excellent canopy, the air, look you, this brave o'erhanging firmament, this majestical roof fretted with golden fire.

—William Shakespeare, *Hamlet*

I wield the flail of the lashing hail,
And whiten the green plains under,
And then again I dissolve it in rain,
And laugh as I pass in thunder.

—Percy Bysshe Shelley, *The Cloud*, 1820

S EVERAL THEORIES OF COSMOGONY attempt to explain the origin of the universe. Without speculating on the validity of any one theory, the following is simply the author's view.

The time: 4,500 million years ago. Before the universe there was time. Only time; otherwise, the vast void held only darkness—everywhere. Overwhelming darkness.

Not dim . . . not murky . . . not shadowy or unlit. Simple nothingness—nothing but darkness, a shade of black so intense we cannot fathom or imag-

---

Information in this chapter adapted from F. R. Spellman, 2009, *The Science of Air*, 2nd ed., Boca Raton, FL: CRC Press.

ine it today. Light had no existence—this was the black of blindness, of burial in the bowels of the earth, the blackness of no other choice.

With time—eons of time—darkness came to a sudden, smashing, shattering, annihilating, scintillating, cataclysmic end—and there was light . . . light everywhere. This new force replaced darkness and lit up the expanse without end, creating a brightness fed by billions of glowing round masses so powerful as to renounce and overcome the darkness that had come before.

With the light was heat-energy that shone and warmed and transformed into mega-mega-mega trillions of super-excited ions, molecules, and atoms—heat of unimaginable proportions, forming gases—gases we don't even know how to describe, how to quantify, let alone how to name. But gases they were—and they were everywhere.

With light, energy, heat, and gases present, the stage was set for the greatest show of all time, anywhere—ever: the formation of the universe.

Over time—time in stretches we cannot imagine, so vast we cannot contemplate them meaningfully—the heat, the light, the energy, and the gases all came together and grew, like an expanding balloon, into one solid glowing mass. But it continued to grow, with the pangs, sweating, and moans accompanying any birthing, until it had reached the point of no return—explosion level. And it did; it exploded with the biggest bang of all time (with the biggest bang hopefully of all time).

The Big Bang sent masses of hot gases in all directions—to the farthest reaches of anything, everything—into the vast, wide, measureless void. Clinging together as they rocketed, soared, and swirled, forming galaxies that gradually settled into their arcs through the void, constantly propelled away from the force of their origin, these masses began their eternal evolution.

Two masses concern us: the sun and earth. Forces well beyond the power of the sun (beyond anything imaginable) stationed this massive gaseous orb approximately 93 million miles from the dense molten core enveloped in cosmic gases and the dust of time that eventually became the insignificant mass we now call Earth.

Distant from the sun, earth's mass began to cool, slowly; the progress was slower than we can imagine, but cool it did. While the dust and gases cooled, earth's inner core, mantle, and crust began to form—no more a quiet or calm evolution than the revolution that cast it into the void had been.

Downright violent was this transformation—the cooling surface only a facade for the internal machinations going on inside, out-gassing from huge, deep destructive vents (we would call them volcanoes today) erupting continuously—never stopping, blasting away, delivering two main ingredients: magma and gas.

The magma worked to form the primitive features of earth's early crust. The gases worked to form earth's initial atmosphere—our point of interest. Without atmosphere, what is there?

About 4 billion years before present, earth's early atmosphere was chemically reducing, consisting primarily of methane, ammonia, water vapor, and hydrogen—for life as we know it today, an inhospitable brew.

Earth's initial atmosphere was not a calm, quiet, quiescent environment; to the contrary, it was an environment best characterized as dynamic—ever changing—where bombardment after bombardment by intense, bond-breaking ultraviolet light, along with intense lightning and radiation from radionuclides, provided energy to bring about chemical reactions that resulted in the production of relatively complicated molecules, including amino acids and sugars (the building blocks of life).

About 3.5 billion years before present, primitive life formed in two radically different theaters: on earth and below the primordial seas near hydrothermal vents that spotted the wavering, water-covered floor.

Initially, on earth's unstable surface, these very primitive life forms derived their energy from fermentation of organic matter formed by chemical and photochemical processes; they then gained the ability to produce organic matter (CH_2O) by photosynthesis.

Thus the stage was set for the massive biochemical transformation that resulted in the production of almost all the atmosphere's O_2.

The O_2 initially produced was quite toxic to primitive life forms. However, much of this oxygen was converted to iron oxides by reaction with soluble iron. This process formed enormous deposits of iron oxides—the existence of which provides convincing evidence for the liberation of O_2 in the primitive atmosphere.

Eventually, enzyme systems developed that enabled organisms to mediate the reaction of waste-product oxygen with oxidizable organic matter in the sea. Later, the mode of waste gradient disposal was utilized by organisms to produce energy by respiration, which is now the mechanism by which non-photosynthetic organisms obtain energy. In time, O_2 accumulated in the atmosphere. In addition to providing an abundant source of oxygen for respiration, the accumulated atmospheric oxygen formed an ozone (O_3) shield—the O_3 shield absorbs bond-rupturing ultraviolet radiation.

With the O_3 shield protecting tissue from destruction by high energy ultraviolet radiation, the earth, although still hostile to life forms we are familiar with, became a much more hospitable environment for life (self-replacing molecules), and life forms were enabled to move from the sea (where they flourished next to the hydrothermal gas vents) to the land. And from that point on to the present, earth's atmosphere became more life-form friendly.

And once formed, it had to be maintained—actually retained. That is, once earth's atmosphere was in place and of composition suitable for life as we know it, it had to be held in place and not just drift willy-nilly out into space.

What it takes for the earth (or any other planet) to hold an atmosphere is a high mass and low temperature. We certainly do not want our atmosphere to be like Mars, Venus, and other planets. Well, some would argue, what is really needed to hold earth's atmosphere in place is gravity. And this is true, of course, but earth's gravity is attributed to its mass. Because of earth's large mass relative to that of the moon, the escape velocity of any particle on earth is 7.1 miles a second, whereas on the moon, the escape velocity is 1.5 miles a second. In simple terms, what this means is that if a particle pointed its nose out to the universe and traveled at 7.1 miles per second it could escape the earth's gravitational hole. The temperature of a planet is important because the higher the temperature, the faster "air" moles travel and collide with one another, and if they are excited enough by the energy input of a high temperature, they might reach escape velocity and leak off into space.

Earth's Thin Skin

Shakespeare likened it to a majestic overhanging roof (constituting the transition between its surface and the vacuum of space); others have likened it to the skin of an apple. Both these descriptions of our atmosphere are fitting, as is its being described as the earth's envelope, veil, or gaseous shroud. The atmosphere is more like the apple skin, however. This thin skin, or layer, contains the life-sustaining oxygen (21%) required by all humans and many other life forms; the carbon dioxide (0.03%) so essential for plant growth; the nitrogen (78%) needed for chemical conversion to plant nutrients; the trace gases such as methane, argon, helium, krypton, neon, xenon, ozone, and hydrogen; and varying amounts of water vapor and airborne particulate matter. Life on earth is supported by this atmosphere, solar energy, and the other planets' magnetic fields. The weight of the major constituents of air are listed in Table 3.1.

Gravity holds about half the weight of a fairly uniform mixture of these gases in the lower 18,000 feet of the atmosphere; approximately 98% of the material in the atmosphere is below 100,000 feet.

Atmospheric pressure varies from 1,000 millibars (mb) at sea level to 10 mb at 100,000 feet. From 100,000 to 200,000 feet, the pressure drops from 9.9 mb to 0.1 mb and so on.

The atmosphere is considered to have a thickness of 40 to 50 miles; however, here we are primarily concerned with the troposphere, the part of the

TABLE 3.1

		Percentage		Molecular Weight		Gram Molecular Weight
N_2	=	78.084	$\times$	28.016	=	21.90
O_2	=	20.946	$\times$	32	=	6.72
Ar	=	0.934	$\times$	39.944	=	0.37
CO_2	=	0.033	$\times$	44.011	=	0.01
						29.00 (approximately)

earth's atmosphere that extends from the surface to a height of about 27,000 feet above the poles, about 36,000 feet in mid latitudes, and about 53,000 feet over the equator. Above the troposphere is the stratosphere, a region that increases in temperature with altitude (the warming is caused by absorption of the sun's radiation by ozone) until it reaches its upper limit of 260,000 feet.

The Troposphere

Extending above earth approximately 27,000 feet, the troposphere is the focus of this text not only because people, plants, animals, and insects live here and depend on this thin layer of gases but also because weather is generated in this region. There is a great deal of mixing in the troposphere. Within the troposphere is the phenomenon of the jet stream. The jet stream is very influential in determining weather patterns. The jet stream undulates snake-like from north to south across North America. If it goes northward, the weather becomes warmer in the south; if it moves southward, the weather becomes warmer in the north. The troposphere begins at ground level and extends 7.5 miles up into the sky where it meets with the second layer called the stratosphere.

Did You Know?

While the gases discussed above are important to maintaining life as we know it on earth, it is water vapor (in conjunction with airborne particles, obviously) that is essential for the stability of earth's ecosystem. This water vapor–particle combination interacts with the global circulation of the atmosphere and produces the world's weather, including clouds and precipitation.

The Stratosphere

The stratosphere begins at the 7.5-mile point and reaches 21.1 miles into the sky. In the rarified air of the stratosphere, the significant gas is ozone (life-protecting ozone, O_3—not to be confused with pollutant ozone), which is produced by the intense ultraviolet radiation from the sun. In quantity, the total amount of ozone in the atmosphere is so small that if it were compressed to a liquid layer over the globe at sea level, it would have a thickness of less than 3/16 of an inch.

In the stratosphere, middle-sized wavelengths of ultraviolet (UV) radiation are absorbed and used when O_1 and O_2 combine to make O_3—ozone. Ozone's ability to absorb UV radiation increases the energy level contained in the stratosphere and increases heat level.

Ozone contained in the stratosphere can also impact (add to) ozone in the troposphere. Normally, the troposphere contains about 20 parts per billion parts of ozone. On occasion, however, via the jet stream, this concentration can increase to 5 to 10 times higher than average.

In our discussion of the earth's atmosphere in this book, the focus is on the troposphere and stratosphere because these two layers directly impact life as we know it and are, or can be, heavily influenced by pollution and its effects.

Did You Know?

The troposphere, stratosphere, mesosphere, and thermosphere act together as a giant safety blanket. They keep the temperature on the earth's surface from dipping to an extreme icy cold that would freeze everything solid or from soaring to blazing heat that would burn up all life.

A Jekyll-and-Hyde View of the Atmosphere

When non-city dwellers look up into that great natural canopy above our heads, they see many features, provided by our world's atmosphere, that we know and enjoy: the blueness and clarity of the sky, the color of a rainbow, the spattering of stars reaching every corner of blackness, the magical colors of a sunset. The air they breathe carries the smell of ocean air, the refreshing breath of clean air after a thunderstorm, and the beauty contained in a snowflake.

But the atmosphere sometimes presents another face—Mr. Hyde's face. The terrible destructiveness of a hurricane, tornado, monsoon, typhoon, or hailstorm; the wearying monotony of winds carrying dusts; and rampaging windstorms carrying fire up a hillside—these are some of the terrifying aspects of the other face of the disturbed atmosphere.

The atmosphere can also present a Hyde-like face whenever man is allowed to pour his filth (pollution) into it: a view of it afforded from patches here and there that are not blocked by buildings; pollution rising from man's enterprises that can mask the stars and make the visible sky a dirty yellow-brown or at best a sickly pale blue.

Fortunately, earth's atmosphere is self-healing. Air cleaning is provided by clouds and the global circulation system that constantly purges the air of pollutants. Only when air pollutants overload nature's way of rejuvenating its systems to their natural state are we faced with the repercussions that can be serious, even life threatening.

Atmospheric Particulate Matter

Along with gases and water vapor, earth's atmosphere is literally a boundless arena for particulate matter of many sizes and types. Atmospheric particulates vary in size from 0.0001 to 10,000 microns. Particulate size and shape have a direct bearing on visibility. For example, a spherical particle in the 0.6-micron range can effectively scatter light in all directions, reducing visibility.

The types of airborne particulates in the atmosphere vary widely, with the largest sizes derived from volcanoes, tornadoes, waterspouts, burning embers from forest fires, seed parachutes, spider webs, pollen, soil particles, and living microbes.

The smaller particles (the ones that scatter light) include fragments of rock, salt and spray, smoke, and particles from forested areas. The largest portion of airborne particulates are invisible. They are formed by the condensation of vapors, chemical reactions, photochemical effects produced by ultraviolet radiation, and ionizing forces that come from radioactivity, cosmic rays, and thunderstorms.

Airborne particulate matter is produced either by mechanical weathering, breakage, and solution or by the vapor-to-condensation-to-crystallization process (typical of particulates from a furnace of a coal-burning power plant).

As you might guess, anything that goes up must eventually come down. This is typical of airborne particulates also. Fallout of particulate matter depends mostly on their size and, less obviously, on their shape, density, weight, airflow, and injection altitude. The residence time of particulate matter also is dependent on the atmosphere's cleanup mechanisms (formation of clouds and precipitation) that work to remove them from their suspended state.

Some large particulates may only be airborne for a matter of seconds or minutes, with intermediate sizes able to stay afloat for hours or days. The finer particulates may stay airborne for a much longer duration: for days, weeks, months, and even years.

Particles play an important role in atmospheric phenomena. For example, particulates provide the nuclei upon which ice particles are formed, cloud condensation forms, and condensation takes place. The most important role airborne particulates play, however, is in cloud formation. Simply put, without clouds life as we know it would be much more difficult, and cloud bursts that eventually erupted would cause such devastation that it is hard to imagine or contemplate.

The situation just described could also result whenever massive forest fires and volcanic action takes place. These events would release a superabundance of cloud condensation nuclei that would overseed the clouds, causing massive precipitation to occur. If natural phenomena such as forest fires and volcanic eruptions can overseed clouds and cause massive precipitation, then what effect would result from man-made pollutants entering the atmosphere at unprecedented levels?

References and Recommended Reading

EPA. 2007. *Air Pollution Control Orientation Course: Air Pollution.* Accessed January 5, 2008, at www.epa.gov/air/oaqps/eog/course422/index.html.

Spellman, F. R., & Whiting, N. 2006. *Environmental Science and Technology: Concepts and Applications,* 2nd ed. Rockville, MD: Government Institutes.

4

Moisture in the Atmosphere

Hath the rain a father? or who hath begotten the drops of dew? Out of whose womb came the ice? and the hoary frost of heaven, who hath gendered it? . . . Can't thou lift up thy voice to the clouds, that abundance of water may cover thee?

—Job 38:28–29, 34

I wandered lonely as a cloud
That floats on high o'er vales and hills,
When all at once I saw a crowd,
A host, of golden daffodils;
Beside the lake, beneath the trees,
Fluttering and dancing in the breeze.

—William Wordsworth, 1804

I sift the snow on the mountains below,
And their great pines groan aghast;
And all the night 'tis my pillow white,
While I sleep in the arms of the blast.

—Percy Bysshe Shelley, *The Cloud*, 1820

Much of this chapter is adapted from F. R. Spellman, 2009, *The Science of Air*, 2nd ed., Boca Raton, FL: CRC Press.

O N A HOT DAY WHEN CLOUDS build up signifying that a storm is imminent, we do not always appreciate what is happening. What is happening?

This cloud buildup actually signals that one of the most vital processes in the atmosphere is occurring: the condensation of water as it is raised to higher levels and cooled within strong updrafts of air created either by convection currents, turbulence, or physical obstacles like mountains. The water originated from the surface—evaporated from the seas, from the soil, or transpired by vegetation. Once within the atmosphere, however, a variety of events combine to convert the water vapor (produced by evaporation) to water droplets. The air must rise and cool to its dew point, of course. At dew point, water condenses around minute airborne particulate matter to make tiny droplets forming clouds—clouds from which precipitation occurs.

Whether created by the sun heating up a hillside, by jet aircraft exhausts, or by factory chimneys, there are actually only 10 major cloud types. Delivering countless millions of tons of moisture from the earth's atmosphere, they form even from the driest desert air containing as little as 0.1% water vapor. They not only provide a visible sign of motion but also indicate change in the atmosphere portending weather conditions that may be expected up to 48 hours ahead. In this chapter we take a brief look at the nature and consequences of these cloud-forming processes.

Cloud Formation

The atmosphere is a highly complex system, and the effects of changes in any single property tend to be transmitted to many other properties. The most profound effect on the atmosphere is the result of alternate heating and cooling of the air; this causes adjustments in relative humidity and buoyancy, which in turn cause condensation, evaporation, and cloud formation.

The temperature structure of the atmosphere (along with other forces that propel the moist air upward) is the main force behind the form and size of clouds. Exactly how does temperature affect atmospheric conditions? For one thing, temperature (i.e., heating and cooling of the surface atmosphere) causes vertical air movements. Let's take a look at what happens when air is heated.

Let's start with a simple parcel of air in contact with the ground. As the ground is heated, the air in contact with it will also warm. This warm air increases in temperature and expands. Remember, gases expand on heating much more than liquids or solids, so this expansion is quite marked. In addition, as the air expands, its density falls (meaning that the same mass of air now occupies a larger volume). You've heard that warm air rises? Because of its lessened density, this parcel of air is now lighter than the surrounding air and tends to rise. Conversely, if the air cools, the opposite occurs—it contracts, its density increases, and it sinks. Actually, alternate heating and cooling are intimately linked with the process of evaporation, condensation, and precipitation.

But how does a cloud actually form? Let's look at another example.

On a sunny day, some patches of ground warm up more quickly than others because of differences in topography (soil and vegetation, etc.). As the surface temperature increases, heat passes to the overlying air. Later, by mid morning, a bulbous mass of warm, moisture-laden air rises from the ground. This mass of air cools as it meets lower atmospheric pressure at higher altitudes. If cooled to its dew point temperature, condensation follows and a small cloud forms. This cloud breaks free from the heated patch of ground and drifts with the wind. If it passes over other rising air masses, it may grow in height. The cloud may encounter a mountain and be forced higher still into the air. Condensation continues as the cloud cools; and if the droplets it holds become too heavy, they fall as rain.

Did You Know?

Clouds play an important role in boundary layer meteorology and air quality. Convective clouds transport pollutants vertically, allowing an exchange of air between the boundary layer and the free troposphere. Cloud droplets formed by heterogeneous nucleation on aerosols grow into rain droplets through condensation, collision, and coalescence. Clouds and precipitation scavenge pollutants from the air. Once inside the cloud or rain water, some compounds dissociate into ions and/or react with one another through aqueous chemistry. Another important role for clouds is the removal of pollutants trapped in rain water and its deposition onto the ground.

Major Cloud Types

There are 10 major cloud types. The clouds were named by Luke Howard (1772–1864), who used Latin descriptive words to distinguish different clouds and to describe the appearance of clouds as seen by an observer on the ground. The descriptive words were based on the cloud's shape and were readily adapted and are still used today.

Stratiform Genera

- Cirrus = hair
- Cirrostratus = high sheets, layers
- Cirrocumulus = high rain
- Altostratus = middle sheets, layers
- Altocumulus = middle piles or heaps
- Stratus = sheets, layers
- Stratocumulus = low piles or heaps
- Nimbostratus = rain sheets, layers

Cumuliform Genera

- Cumulus = piles or heaps
- Cumulonimbus = piles or heaps, rain

Further classification identifies clouds by height of cloud base. For example, cloud names containing the prefix "cir-," as in cirrus clouds, are located at high levels; while cloud names with the prefix "alto-," as in altostratus, are found at middle levels. This module introduces several cloud groups. The first three groups are identified based upon their height above the ground. The fourth group consists of vertically developed clouds, while the final group consists of a collection of miscellaneous cloud types. In naming a cloud, state its elevation first, then its cloud type. Cloud bases and types are measured vertically.

Did You Know?

The deeper the color of the clouds at sunset, the lower the cloud; the whiter, the higher.

TABLE 4.1
Summary of Components of Cloud Classification System

Latin Root	Translation	Example
Cumulus	Heaped/puffy	Fair weather cumulus
Stratus	Layered	Altostratus
Cirrus	Curl of hair/wispy	Cirrus
Nimbus	Rain	Cumulonimbus

Let's take a closer look at each of these cloud types.

A *stratus* cloud is a featureless, gray, low-level cloud. Its base may obscure hilltops or occasionally extend right down to the ground, and because of its low altitude, it appears to move very rapidly on breezy days. Stratus can produce drizzle or snow, particularly over hills, and may occur in huge sheets covering several thousand miles.

Cumulus clouds also seem to scurry across the sky, reflecting their low altitude. These small, dense, white, fluffy, flat-based clouds are typically short lived, lasting no more than 10 to 15 minutes before dispersing. They are typically formed on sunny days, when localized convection currents are set up: These currents can form over factories or even brush fires, which may produce their own clouds.

Cumulus may expand into low-lying, horizontally layered, massive *strato-cumulus*, or into extremely dense, vertically developed, giant *cumulonimbus* with a relatively hazy outline and a glaciated top that can reach 7 miles in diameter. These clouds typically form on summer afternoons; their high, flattened tops contain ice, which may fall to the ground in the form of heavy showers of rain or hail.

Rising to middle altitudes, the bluish-gray layered *altostratus* and rounded, fleecy, whitish-gray *altocumulus* appear to move slowly because of their greater distance from the observer.

Cirrus clouds (*cirrus,* meaning tuft of hair) are made up of white narrow bands of thin, fleecy parts and are relatively common over northern Europe; they generally ride the jet stream rapidly across the sky.

Cirrocumulus are high-altitude clouds composed of a series of small, regularly arranged cloudlets in the form of ripples or grains; they are often present with cirrus clouds in small amounts. *Cirrostratus* are high-altitude, thin, hazy clouds, usually covering the sky and giving a halo effect surrounding the sun or moon.

Did You Know?

Clouds whose names incorporate the word "nimbus" or the prefix "nimbo-" are clouds from which precipitation is falling.

Moisture in the Atmosphere

Let's summarize the information related to how moisture accumulates in and precipitates from the atmosphere. The process of evaporation (converting moisture into vapor) supplies moisture into the lower atmosphere. The prevailing winds then circulate the moisture and mix it with drier air elsewhere.

Water vapor is only the first stage of the precipitation cycle; the vapor must be converted into liquid form. This is usually achieved by cooling, either rapidly, as in convection, or slowly, as in cyclonic storms. Mountains also cause uplift, but the rate will depend upon their height and shape and the direction of the wind.

To actually produce precipitation, the cloud droplets must become large enough to reach the ground without evaporating. The cloud must possess the right physical properties to enable the droplets to grow.

If the cloud lasts long enough for growth to take place, then precipitation will usually occur. Precipitation results from a delicate balance of counteracting forces, some leading to droplet growth and others to droplet destruction.

Did You Know?

Contrails are clouds formed around the small particles (aerosols) that are in aircraft exhaust. When these persist after the passage of the plane, they are indeed clouds and are of great interest to researchers. Under the right conditions, clouds initiated by passing aircraft can spread with time to cover the whole sky.

References and Recommended Reading

Ahrens, D. 1994. *Meteorology Today: An Introduction to Weather, Climate and the Environment*, 5th ed. West Publishing Company.

EPA. 2007. *Air Pollution Control—Atmosphere*. Accessed December 28, 2007, at www.epa.gov/air/oaqps/eog/course422/index.html.

NASA. 2008. *Observing Cloud Type*. Washington, DC: National Aeronautics and Space Administration.

NOAA. 2007. *Cloud Types*. Accessed December 29, 2007, at www.gfdl.NOAA.gov/~io/weather/clouds.html.

Spellman, F. R., & Whiting, N., 2006. *Environmental Science & Technology: Concepts and Applications*, 2nd ed. Rockville, MD: Government Institutes.

5

Precipitation and Evapotranspiration

BECAUSE IT DETERMINES THE INTENSITY and distribution of many of the processes operating within the system, precipitation is one of the most important regulators of the hydrological cycle. The rate of evapotranspiration is closely related to precipitation and thus is also an integral part of the hydrological cycle.

The Rainy Day
The day is cold, and dark, and dreary;
It rains, and the wind is never weary;
The vine still clings to the moldering wall,
But at every gust the dead leaves fall,
And the day is dark and dreary.

My life is cold, and dark, and dreary;
It rains, and the wind is never weary;
My thoughts still cling to the moldering Past,
But the hopes of youth fall thick in the blast
And the days are dark and dreary.

Be still, sad heart! And cease repining;
Behind the clouds is the sun still shining;
Thy fate is the common fate of all,
Into each life some rain must fall,
Some days must be dark and dreary.

—Henry Wadsworth Longfellow

Sublime on the towers of my skiey bowers,
Lightning my pilot sits;
In a cavern under is fettered the thunder,
It struggles and howls at fits;
Over Earth and Ocean, with gentle motion,
This pilot is guiding me,
Lured by the love of the genii that move
In the depths of the purple sea;
Over the rills, and the crags, and the hills,
Over the lakes and the plains,
Wherever he dream, under mountain or stream,
The Spirit he loves remains;
And I all the while bask in Heaven's blue smile,
Whilst he is dissolving in rains.

—Percy Bysshe Shelley, *The Cloud*, 1820

(Much of this introductory section is adapted from F. R. Spellman, 2007, *The Science of Air*, 2nd ed., Boca Raton, FL: CRC Press.) The principal actions brought on by weather systems that affect land and sea and the humans, animals, and vegetation thereon are winds and precipitation. The latter comes in a variety of forms as discussed here. Most weather of consequence to people occurs in storms. These may be local in origin but more commonly are carried to locations in wide areas along pathways followed by active air masses consisting of highs and lows. The key ingredient in storms is water, either as a liquid or as a vapor. The vapor acts like a gas and thus contributes to the total pressure of the atmosphere, making up a small but vital fraction of the total (NASA, 2008).

Precipitation is found in a variety of forms. Which form actually reaches the ground depends upon many factors: for example, atmospheric moisture content, surface temperature, intensity of updrafts, and method and rate of cooling.

Water vapor in the air will vary in amount depending on sources, quantities, processes involved, and air temperature. Heat, mainly as solar irradiation but with some contributed by the earth and human activity and some from change-of-state processes, will cause some water molecules either in water bodies (oceans, lakes, rivers) or in soils to be excited thermally and escape from their sources. This is called evaporation; if water is released from trees and other vegetation the process is known as evapotranspiration. The evaporated water, or moisture, that enters the air is responsible for a state called humidity. Absolute humidity is the weight of water vapor contained in a given volume of air. The mixing ratio refers to the mass of the water vapor within a given mass of dry air. At any particular temperature, the maximum amount of water vapor that can be contained is limited to some amount; when that amount is reached, the air is said to be saturated for that temperature. If less than the maximum amount is present, then the property of air that indicates this is its relative humidity (RH), defined as the actual water vapor amount compared to the saturation amount at the given temperature; this is usually expressed as a percentage. RH also indicates how much moisture the air can hold above its stated level that, after attaining, could lead to rain.

When a parcel of air attains or exceeds RH = 100%, condensation will occur and water in some state will begin to organize as some type of precipitation. One familiar form is dew, which occurs when the saturation temperature or some quantity of moisture reaches a temperature at the surface at which condensation sets in, leaving the moisture to coat the ground (especially obvious on lawns).

The term *dew point* has a more general use, being that temperature at which an air parcel must be cooled to become saturated. Dew frequently forms when the current air mass contains excessive moisture after a period of rain but the air is now clear; the dew precipitates out to coat the surface (noticeable on vegetation). Ground fog is a variant in which lower temperatures bring on condensation within the near surface air as well as the ground.

The other types of precipitation are listed in Table 5.1 along with descriptive characteristics related to each type.

Evaporation and transpiration are complex processes that return moisture to the atmosphere. The rate of evapotranspiration depends largely on two factors: (1) how saturated (moist) the ground is and (2) the capacity of the atmosphere to absorb the moisture. In this chapter we discuss the factors responsible for both precipitation and evapotranspiration.

Phases of Precipitation

Precipitation falls in various forms, or phases, that can be subdivided into the following categories (METAR, 2011).

Liquid Precipitation

- Drizzle
- Rain

TABLE 5.1
Types of Precipitation

Type	Approximate Size	State of Water	Description
Mist	0.005 to 0.05 mm	Liquid	Droplets large enough to be felt on face when air is moving 1 meter/second; associated with stratus clouds.
Drizzle	Less than 0.5 mm	Liquid	Small uniform drops that fall from stratus clouds, generally for several hours.
Rain	0.5 to 5 mm	Liquid	Generally produced by nimbostratus or cumulonimbus clouds. When heavy, size can be highly variable from one place to another.
Sleet	0.5 to 5 mm	Solid	Small, spherical to lumpy ice particles that form when raindrops freeze while falling through a layer of sub-freezing air. Because the ice particles are small, any damage is generally minor. Sleet can make travel hazardous.
Glaze	Layers 1 mm to 2 cm thick	Solid	Produced when supercooled raindrops freeze on contact with solid objects. Glaze can form a thick covering of ice having sufficient weight to seriously damage trees and power lines.
Rime	Variable accumulation	Solid	Deposits usually consist of ice feathers that point into the wind. These delicate frostlike accumulations form as supercooled cloud or fog droplets encounter objects and freeze on contact.
Snow	1 mm to 2 cm	Solid	The crystalline nature of snow allows it to assume many shapes, including six-sided crystals, plates, and needles. Produced in supercooled clouds where water vapor is deposited as crystals that remain frozen during their descent.
Hail	5 mm or larger	Solid	Precipitation in the form of hard, rounded pellets or irregular lumps of ice. Produced in large convective, cumulonimbus clouds, where frozen ice particles and supercooled water coexist.
Graupel	2 mm to 5 mm	Solid	Sometimes called "soft hail," graupel forms as rime collects on snow crystals to produce irregular masses of "soft" ice. Because these particles are softer than hailstones, they normally flatten out upon impact.

Source: NASA, 2008.

Freezing Precipitation

- Freezing drizzle
- Freezing rain

Frozen Precipitation

- Snow
- Snow grains
- Ice pellets
- Hail
- Snow pellets/graupel
- Ice crystals

Precipitation

As mentioned, if all the essentials are present, precipitation occurs when the dew point is reached. However, it was also pointed out that it is quite possible for an air mass or cloud containing water vapor to be cooled below the dew point without precipitation occurring. In this state the air mass is said to be *supercooled.*

How, then, are droplets of water formed? Water droplets form around microscopic foreign particles already present in the air. These particles on which the droplets form are called *hygroscopic nuclei.* They are present in the air primarily in the form of dust, salt from sea water evaporation, and combustion residue. These foreign particles initiate the formation of droplets that eventually fall as precipitation. To have precipitation, larger droplets or drops must form. This may be brought about by two processes: (1) coalescence (collision) or (2) the Bergeron process.

Coalescence

Simply put, *coalescence* is the fusing together of smaller droplets into larger ones. In meteorology, its role is crucial in the process by which water droplets in a cloud collide and come together to form raindrops (NOAA, 2011). The variation in the size of the droplets has a direct bearing on the efficiency of this process. Raindrops come in different sizes and can reach diameters up to 7 mm. Having larger droplets greatly enhances the coalescence process; that is, when the droplets become too large to be sustained on the air currents, they begin to fall as rain.

But what actually goes on inside a cloud to cause rain to fall? To answer this question, we must take a look inside a cloud to see exactly what processes occur to make rain—rain that actually falls as rain. Rainmaking is based on the essentials of the Bergeron process.

Bergeron Process

Named after the Swedish meteorologist who suggested it, the *Bergeron process* is probably the more important process for the initiation of precipitation. It is defined as the process by which ice crystals in a cloud grow at the expense of supercooled liquid water droplets (NOAA, 2011). To gain understanding on how the Bergeron process works, let's look at what actually goes on inside a cloud to cause rain.

Within a cloud made up entirely of water droplets, there will be a variety of droplet sizes. The air will be rising within the cloud anywhere from 10 to 20 cm per second (depending on the type of cloud). As the air rises, the drops become larger through collision and coalescence; many will reach drizzle size. Then the updraft intensifies up to 50 cm per second (and more), which reduces the downward movement of the drops, allowing them more time to become even larger. When the cloud becomes approximately 1 km deep, small raindrops of 700 μm diameter are formed.

The droplets, because of their small size, do not freeze immediately, even when the temperatures fall below 0°C. Instead, the droplets remain unfrozen in a supercooled state. However, when the temperature drops as low as –10°C, ice crystals may start to develop among the water droplets. This mixture of water and ice would not be particularly important but for a peculiar characteristic or property of water. Therefore at –10°C, air saturated with respect to liquid water is supersaturated relative to ice by 10% and at –20°C by 21%. Thus ice crystals in the cloud tend to grow and become heavier at the expense of the water droplets.

Eventually, the ice crystals sink to the lower levels of the cloud where temperatures are only just below freezing. When this occurs, they tend to combine (the supercooled droplets of water act as an adhesive) and form snowflakes. When the snowflakes melt, the resulting water drops may grow further by collision with cloud droplets before they reach the ground as rain. The actual rate at which water vapor is converted to raindrops depends on three main factors: (1) the rate of ice crystal growth, (2) supercooled vapor, and (3) the strength of the updrafts (mixing) in the cloud.

Types of Precipitation

We stated that in order for precipitation to occur, water vapor must condense, which occurs when water vapor ascends and cools. Three mechanisms by which air rises, allowing for precipitation to occur, are convectional, orographic, and frontal.

Convectional Precipitation

Convectional precipitation is the spontaneous rising of moist air due to instability caused by surface heating of the air at the ground level. If enough heating occurs, the mass of air becomes warmer and lighter than the air in the surrounding environment, and just like a hot air balloon, it begins to rise, expand, and cool. When sufficient cooling has taken place, saturation occurs from precipitation (NOAA, 2011). This type of precipitation is usually associated with thunderstorms and occurs in the summer because localized heating is required to initiate the convection cycle. We have discussed that upward-growing clouds are associated with convection. Since the updrafts (commonly called a "thermal") are usually strong, cooling of the air is rapid and lots of water can be condensed quickly, usually confined to a local area, and a sudden summer downpour may occur as a result. Convectional thunderstorm clouds are also described as supercells.

Convective thunderstorms are the most common type of atmospheric instability that produces lightning followed by thunder. Lightning is one of the most spectacular phenomena witnessed in storms.

Did You Know?

A lightning bolt can attain an electric potential up to 30 million volts and current as much as 10,000 amps. It can cause air temperatures to reach 10,000°C. But a bolt's duration is extremely short (fractions of a second). Although a bolt can kill people it hits, most can survive.

Orographic Precipitation

Orographic precipitation is a straightforward process, characteristic of mountainous regions; almost all mountain areas are wetter than the surrounding lowlands. This type of precipitation arises when masses of air are forced to rise over a mountain or mountain range. The wind, blowing along the surface of the earth, ascends along topographic variations. Where air meets this extensive barrier, it is forced to rise. This ascending wind usually gives rise to cooling and encourages condensation and thus orographic precipitation on the windward side and drier air on the leeward side of the mountain range.

Frontal Precipitation

Frontal precipitation results when two different fronts (or the boundary between two air masses characterized by varying degrees of precipitation), at different temperatures, meet. The warm air mass (since it is lighter) moves up and over the colder air mass. The cooling is usually less rapid than in the vertical convection process because the warm air mass moves up at an angle, more of a horizontal motion.

Evapotranspiration

Another important part or process of the hydrological cycle (though it is often neglected because it can rarely be seen) is *evapotranspiration*. More complex than precipitation, evaporation and transpiration is a land-atmosphere interface process whereby a major flow of moisture is transferred from ground level to the atmosphere. It returns moisture to the air, replenishing that lost by precipitation, and it also takes part in the global transfer of energy. The rate of evapotranspiration depends largely on two factors: (1) how moist the ground is and (2) the capacity of the atmosphere to absorb the moisture. Therefore the greatest rates are over the tropical oceans, where moisture is always available and the long hours of sunshine and steady trade winds evaporate vast quantities of water.

Just how much moisture is returned to the atmosphere via transpiration? In answering this question, Table 5.2 makes clear, for example, that in the United States alone, about two-thirds of the average rainfall over the U.S. mainland is returned via evaporation and transpiration.

Evaporation

Evaporation is the process by which a liquid is converted into a gaseous state. Evaporation takes place (except when air reaches saturation at 100% humidity) almost on a continuous basis. It involves the movement of individual water molecules from the surface of earth into the atmosphere, a process occurring whenever a vapor pressure gradient exists from the surface to the

TABLE 5.2
Water Balance in the United States (in BGD, Billion Gallons per Day)

Precipitation	4,200
Evaporation and transpiration	3,000
Runoff	1,250
Withdrawal	310
Irrigation	142
Industry (utility cooling water)	142
Municipal	26
Consumed (irrigation loss)	90
Returned to streams	220

Source: National Academy of Sciences, 1962.

air (i.e., whenever the humidity of the atmosphere is less than that of the ground). Evaporation also requires energy (derived from the sun or from sensible heat from the atmosphere or ground): 2.48×10^6 joules to evaporate each kilogram of water at 10°C.

Transpiration

A related process, *transpiration* is the loss of water from a plant by evaporation. Most water is lost from the leaves through pores known as stomata, whose primary function is to allow gas exchange between the plant's internal tissues and the atmosphere. Transpiration from the leaf surfaces causes a continuous upward flow of water from the roots via the xylem, which is known as the transpiration stream.

Transpiration occurs mainly by day, when the stomata open up under the influence of sunlight. Acting as evaporators, they expose the pure moisture (the plant's equivalent of perspiration) in the leaves to the atmosphere. If the vapor pressure of the air is less than that in the leaf cells, the water is transpired.

As you might guess, because of transpiration, far more water passes through a plant than is needed for growth. In fact, only about 1% or so is actually used in plant growth. Nevertheless, the excess movement of moisture through the plant is important to the plant because the water acts as a solvent, transporting vital nutrients from the soil into the roots and carrying them through the cells of the plant. Obviously, without this vital process plants would die.

The Process of Evapotranspiration

Although evapotranspiration plays a vital role in cycling water over earth's land masses, it is seldom appreciated. In the first place, distinguishing between evaporation and transpiration is often difficult. Both processes tend to be operating together, so the two are normally combined to give the composite term *evapotranspiration.*

Governed primarily by atmospheric conditions, energy is needed to power the process. Wind also plays an important role, acting to mix the water molecules with the air and transport them away from the surface. The primary limiting factor in the process is lack of moisture at the surface (soil is dry). Evaporation can continue only so long as there is a vapor pressure gradient between the ground and the air.

References and Recommended Reading

METAR. 2011. *METAR Conversion Card.* National Weather Service. Accessed July 1, 2011, at www.nws.noas.gov/oso/oso1/oso12/document/guide.shtml.

NASA. 2008. *Observing Cloud Type.* Washington, DC: National Aeronautics and Space Administration.

National Academy of Sciences. 1962. *Water Balance in the U.S.* National Research Council Publication 100-B.

NOAA. 2011. *National Weather Service Glossary.* Accessed July 1, 2011, at www.weather.gov/glossary/.

Shipman, J. T., Adams, J. L., & Wilson, J. D. 1987. *An Introduction to Physical Science.* Lexington, MA: D.C. Heath and Company.

Spellman, F. R. 2007. *The Science of Water,* 2nd ed. Boca Raton, FL: CRC Press.

USGS. 2008. *The Water Cycle: Evapotranspiration.* Accessed January 7, 2008, at http://ga.water.usgs.gov/edu/watercycleevapotranspiration.html.

6

The Seasons

Joys of Spring
The climbing sun again was wakening the world
And laughing at the wreck of frigid winter's trade. . . .
Summer Toils
Hail everchanging world, with May days come and gone.
Hail men, with the advent of sunny summertime. . . .
Autumn Wealth
Look yon, the sun again is rolling down the sky:
Each day shows less and less of its majestic rays,
Stretches out more and more the shadows of all things,
And in a greater haste descends beyond the hills. . . .
Winter Cares
Look yon! What great, fierce fangs the beast of winter bares!
What sullen northern winds roll here to harry us!
On the grim lakes and ponds translucent windows form,
Like shinning mirrors fashioned by a glazier's hand.
The pools, where swam the fish and leaped and dived the frogs,
Because of winter's threats, have put on armor plate,
So now in dark retreats drowse all the water folk.

—Kristijonas Donelaitis, 1818

Bathed in such beauty, watching the expressions ever varying on the faces of the mountains, watching the stars, which here have a glory that the lowlander never dreams of, watching the circling seasons, listening to the songs of the waters and winds and birds, would be endless pleasure. And what glorious cloud-lands I would see, storms and calms, a new heaven and a new earth every day.

—John Muir, 1911

Why Do We Have the Seasons?

(INFORMATION IN THIS SECTION is from the National Weather Service 2011: *Why Do We Have the Seasons?* Accessed July 1, 2011, at www.wrh. noaa.gov/fgz/science/season.php?who=fgz.) Simply, it is the tilt of the earth's spin axis that causes the seasons; the earth spins on its axis, producing night and day. It also moves about the sun in an elliptical (elongated circle) orbit that requires about 365 ¼ days to complete (see Figure 6.1). Again, the earth's spin axis is tilted with respect to its orbital plane. This is what causes the seasons. When the earth's axis points towards the sun, it is summer for that hemisphere. When the earth's axis points away, winter can be expected. Since the tilt of the axis is 23.5°, the north pole never points directly at the sun; but on the summer solstice it points as close as it can, and on the winter solstice, as far as it can. Midway between these two times, in spring and autumn, the spin axis of the earth points 90° away from the sun. This means that on this date, day and night have about the same length: 12 hours each, more or less.

Why should this tilt of the earth's axis matter to our climate? To understand this, take a piece of paper and a flashlight. Shine the light from the flashlight straight onto the paper, so you see an illuminated circle. All the light from the flashlight is in that circle. Now slowly tilt the paper, so that the circle elongates into an ellipse spread out over more paper. The density of

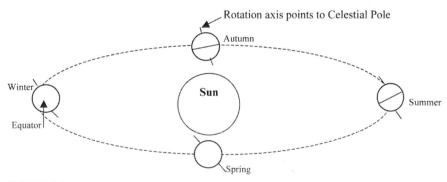

FIGURE 6.1
The earth's rotation axis always points to celestial poles.

light drops. In other words, the amount of light per square centimeter drops (the number of square centimeters increases, while the total amount of light stays the same).

The same is true on earth. When the sun is overhead, the light is falling straight on you, and so more light (and more heat) hits each square centimeter of the ground. When the sun is lower in the sky, the light gets more spread out over the surface of the earth, and less heat (per square centimeter) can be absorbed. Since the earth's axis is tilted, the sun is higher when you are on the part of the earth where the axis points toward the sun and lower on the part of the earth where the axis points away from the sun.

For the northern hemisphere, the axis points most toward the sun in June (specifically around June 21) and away from the sun around December 21. This corresponds to the winter and summer solstice (solstice is Latin for "the sun stands"), or the midpoints of winter and summer. For the southern hemisphere, this is reversed.

For both hemispheres, the earth is 90° away from the sun around March 21 and then again around September 21. This corresponds to the fall and spring equinox (equinox is Latin for "equal night"). Every place in the world has about 12 hours of daylight and 12 hours of night.

Did You Know?

To find out how high the sun is at noontime, do the following:

1. Find the latitude of the place in question.
2. Find the latitude of the place where the sun is exactly overhead.
3. Determine how far it is in degrees from the first place to the second place.
4. Subtract this distance (in degrees) from 90°.

Example: (June 21)

1. Location: latitude 37°
2. Tropic of Cancer: 23.5°
3. Difference: 13.5°
4. 90° − 13.5° = 76.5°
5. This is the highest point the sun can reach over latitude 37°
6. The lowest point sun can reach over latitude 76.5°

The equations is as follows:

$$\frac{-47.0°}{29.5°}$$

Why Are Sunrise and Sunset Not Exactly 12 Hours Apart on the Equinox?

Day and night are not exactly of equal length at the time of the March and September equinoxes. The dates on which day and night are each 12 hours occur a few days before and after the equinoxes. The specific dates for this occurrence are different for different latitudes.

On the day of the equinox, the geometric center of the sun's disk crosses the equator, and this point is above the horizon for 12 hours everywhere on the earth. However, the sun is not simply a geometric point. Sunrise is defined as the instant when the leading edge of the sun's disk becomes visible on the horizon, whereas sunset is the instant when the trailing edge of the disk disappears below the horizon. At these times, the center of the disk is already below the horizon. Furthermore, atmospheric refraction (or bending) of the sun's rays causes the sun's disk to appear higher in the sky than it would if the earth had no atmosphere. Thus, in the morning, the upper edge of the disk is visible for several minutes before the geometric edge of the disk reaches the horizon. Similarly, in the evening, the upper edge of the disk disappears several minutes after the geometric disk has passed below the horizon.

For observers within a couple of degrees of the equator, the period from sunrise to sunset is always several minutes longer than the night. At higher latitudes in the northern hemisphere, the date of equal day and night occurs before the March equinox. Daytime continues to be longer than nighttime until after the September equinox. In the southern hemisphere, the dates of equal day and night occur before the September equinox and after the March equinox.

Table 6.1 shows the dates and times for the equinoxes and solstices for 2011 and 2012. Times listed are in Mountain Standard Time.

Heat Island Effect in Large Cities

The mechanisms that lead to the creation of an urban heat island (UHI; see Figure 6.2) have been studied for nearly a century and are well understood. As the sun rises, it warms the buildings quickly because its rays are perpendicular to them, and then the building radiates that heat to the city. As the sun is set-

TABLE 6.1

Year	Spring Equinox	Summer Solstice	Fall Equinox	Winter Solstice
2011	Mar 20—4:21 pm	June 21—10:16 am	Sept 23—2:04 am	Dec 21—10:30 pm
2012	Mar 19—10:14 pm	June 20—4:09 pm	Sept 22—7:49 am	Dec 21—4:11 pm

ting, the buildings also get full benefit (instead of slant benefit) of the sun's rays, absorbing the energy to later radiate it.

In 2008, the National Weather Service (NWS) published a report related to urban heat island (UHI) effect on rainfall variability across the Phoenix, Arizona, Metropolitan Area (PMA) during the monsoon season. According to the city of Phoenix's website, the population of Phoenix in 1881 was roughly 2,500. Today, the population of the PMA is near 4.2 million. In 2030, the population is projected to be 7.3 million (Maricopa Association of Governments, 2005). This large and mostly rapid growth has resulted in the creation of a significant urban heat island (UHI). But has the development of the UHI led to a change in convective (i.e., thunderstorm) precipitation patterns across the PMA during the monsoon season? To date, no research has conclusively proven that the UHI has altered precipitation patterns across the PMA. However, a growing body of research exists that supports the notion that the PMA is enhancing precipitation in downwind (to the northeast)

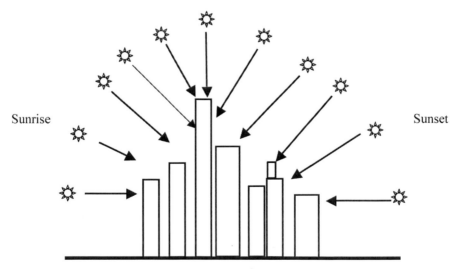

FIGURE 6.2
The sun continuously heats, creating a heat island.

areas. Data present in the NWS 2008 report illustrate that there are no coherent patterns in precipitation across the PMA that cannot be explained by topographic influences.

As stated, like many other large cities, the PMA has a substantial UHI. As early as 1821, urban effects on the local climate in Phoenix, Arizona, were recognized. Analyzing winter-time minimum temperatures in the Salt River Valley, Gordon (1921) found that the weather station in Phoenix, which at the time was roughly 2% of its current population, was warmer than expected for a low-valley location. He noted that the Riverside Nursery is protected on the north by both the diversion of the cold airsteam and by the city of Phoenix itself, with its warming influence, likely marking him the first scientist to describe the Phoenix UHI. Since that time, the population has increased nearly 40-fold, resulting in an intensification and expansion of the UHI.

The effects of the UHI are most pronounced during the summer (June–July–August) months. In a simple comparison, average high and low temperatures from Maricopa, Arizona, and Phoenix Sky Harbor International Airport were compared. Little difference exists between the high temperatures from 1961 through 2007; however, low temperatures have trended warmer at Sky Harbor as compared to Maricopa, indicative of the existence of a strengthening UHI. Similar results have been found by Brazel and colleagues (2000), who compared temperature observations at several urban stations in the PMA (Phoenix, Mesa, and Tempe) to a rural station outside the PMA (Sacation). The average minimum May temperature at the urban locations displayed an apparent upward increase through the time series (1910 through 2000), which was attributed to the urban growth of the region. By the end of the series, there was a +4 to 7°C (7 to 13°F) difference between the urban minimum temperatures and the rural minimum temperatures. Interestingly, no upward trend was found in the maximum temperatures, and in fact, the urban temperature was consistently 0 to 2°C (0 to 4°F) cooler than the rural station. The hypothesized reason for this is the oasis effect, where solar energy is expended on evaporation water from an unnaturally moist surface in the urban areas due to irrigational activities.

Anthropogenic (Human-Caused) Effects on Precipitation

The effects of urban heat islands (UHIs) on rainfall patterns have been anecdotally noted since at least the early 20th century. Horton (1921) described the tendency for cities to be thunderstorm-breeding spots, places that he observed as being favorable for thunderstorm development. While the notion that cities can have a direct impact on precipitation patterns has existed since the early 20th century, it was not until the later 1960s that the topic gained widespread interest. In an analysis of precipitation variables in and around the Chicago, Illinois, area, Changnon (1968) described a statistical anomaly that existed downwind of Chicago, near the town of La Porte, Indiana. Compared to other nearby stations in the dataset, La Porte had experienced a 31% increase in annual precipitation, a 28% increase in warm-season precipitation, a 34% increase in the number of heavy rain days, and a 38% increase in the frequency of thunderstorms. Through an exhaustive analysis of the data, the author was able to eliminate a number of potential sources of error and concluded that the "La Porte anomaly" was a reality. The source of the anomaly was attributed to an increase in cloud condensation nuclei (CCN) released by an industrial complex that lay upwind of La Porte.

There are generally four agreed-upon human-caused mechanisms by which urban areas alter nearby precipitation fields. The first is the addition of sensible heat due to the change in thermal properties associated with the transformation of the landscape (Lowry, 1998; Oke, 2005). This often results in the formation of thermal updrafts over urban areas and is indicative of the urban heat island. Second is the addition of latent heat to the lower atmosphere through anthropogenic activities such as irrigation (Diem and Brown, 2003; Dixon and Mote, 2003). Third is an increase in surface convergence due to an increase in surface roughness associated with urban structures (Loose and Bornstein, 1977; Bornstein and Lin, 2003). The fourth mechanism is pollution. Pollution particles, such as sulfates, can act as cloud condensation nuclei (CCN) that can either promote or retard the growth of water droplets, depending on the structure and physical/chemical properties of the CCN (Gatz, 1974; Salby, 1996; Rosenfeld, 2000).

Since the discovery of the La Porte anomaly, many studies have been published that attempt to quantify to what extent cities influence local rainfall, several of which are specific to central Arizona. Diem and Brown (2003) acquired precipitation data from several widely scattered rain gauges in central Arizona. A trend analysis of the data, from 1950 through 2000, yielded an increase in precipitation at a handful of stations in the Lower Verde River basin. Similar results were found by Shepherd (2006) utilizing a longer dataset bifurcated into two periods, a pre-urban (1900–1950) and post-urban (1950–2000). Satellite-based RADAR data were also used, with similar results to previous research. The common theme to this research was an increase of precipitation downwind (to the northeast) of the Phoenix Metropolitan Area (PMA). No research studies have been published in the scientific community that specifically address the alteration of precipitation across the PMA itself.

The rapidly growing PMA represents an interesting yet complex area to study anthropogenic effects on precipitation (AEP). The topography of the PMA is generally flat, but some of the most complex topography in the state resides just 50 to 120 km (30 to 75 miles) to the northeast of downtown Phoenix. The range of mountainous terrain in central Arizona, known as the Mogollon Rim, is a favored area for convective development during the summer and is one of the most active convective areas in the United States (Changnon, 1988). Precipitation amounts during the monsoon season are greatest along and to the southwest of the Mogollon Rim. In addition, Arizona has one of the most unique diurnal evolutions of convection in the United States. Using lightning data, King and Balling (1994) found that peak lightning activity was confined to the Mogollon Rim and complex terrain of southeast Arizona during the mid-afternoon hours. A very late (near midnight) maximum was found over central Arizona, including the PMA. This type of late-night maximum is found in only one other locale in the country (north-central Great Plains). While the precise reasoning for this extraordinary pattern is unknown, it is likely due to a combination of airflow drainage from the mountains into the PMA (Brazel et al., 2005) and thunderstorm convergence. It has been well documented by several researchers that convection will often propagate to the west or southwest from the Mogollon Rim while convection propagates west or northwest from the mountainous areas of southeast Arizona, thus converging over the PMA (Watson et al., 1994; McCollum et al., 1995). This process generally causes the greatest precipitation amounts around the PMA, clockwise from northwest to south, with the relative maximum to the northeast.

References and Recommended Reading

Borstein, R., and Lin, Q. 2003. "Urban Heat Islands and Summertime Convective Thunderstorms in Atlanta: Three Case Studies." *Atmospheric Environment*, 34, 507–516.

Brazel, A., Fernando, H. J. S., Hunt, J. C. R., Selover, N., Hedquist, B., & Pardyjak, E. 2005. "Evening Transition Observations in Phoenix, Arizona." *Journal of Applied Meteorology*, 44, 99–112.

Brazel, A., Selover, N., Vose, R., & Heisler, G. 2000. "The Tale of Two Climates: Baltimore and Phoenix Urban LTER Sites." *Climate Research*, 15, 123–135.

Changnon Jr., S. A., 1968. "The La Porte Weather Anomaly, Fact or Fiction?" *Bulletin American Meteorology Society*, 49, 4–11.

Changnon Jr., S. A., 1988. "Climatography of Thunder Events in the Conterminous United States, Part II: Spatial Analysis." *Journal of Climate*, 1, 399–405.

Diem, J. E., & Brown, D. P. 2003. "Anthropogenic Impacts on Summer Precipitation in Central Arizona, USA." *The Professional Geographer*, 55, 343–355.

Dixon, P. G., & Mote, T. L. 2003. "Patterns and Causes of Atlanta's Urban Heat Island-Initiated Precipitation." *Journal of Applied Meteorology*, 42, 1273–1283.

Gatz, D. R. 1974. "METROMEX: Air and Rain Chemistry Analysis." *Bulletin American Meteorolology Society,* 55, 92–93.

Gordon, J. H. 1921. "Temperature Survey of the Salt River Valley, Arizona." *Monthly Weather Review*, 49, 271–275.

Horton, R. E. 1921. "Thunderstorm-Breeding Spots." *Monthly Weather Review*, 49, 193.

King, T. S., & Balling Jr., R. C. 1994. "Diurnal Variations in Arizona Monsoon Lightning Data." *Monthly Weather Review*, 122, 1659–1664.

Loose, T., & Bornstein, R.D. 1977. "Observations of Mesoscale Effects on Frontal Movement through an Urban Area." *Monthly Weather Review*, 105, 563–571.

Lowry, W. P. 1998. "Urban Effects on Precipitation Amount." *Progress in Physical Geography*, 22(4), 477–520.

Maricopa Association of Governments. 2005. "Regional Report: A Resource for Policy Makers in the Maricopa Region, AZ."

McCollum, D. M., Maddox, R. A., & Howard, K. W. 1995. "Case Study of a Severe Mesoscale Convective System in Central Arizona." *Weather and Forecasting*, 10, 643–665.

Oke, T. R. 2005. "Towards Better Scientific Communication in Urban Climate." *Theoretical and Applied Climatology*. DOI 10.1007/s00704-005-0153-0.

Rosenfeld, D. 2000. "Suppression of Rain and Snow by Urban and Industrial Air Pollution." *Science*, 287, 1793–1796.

Salby, M. L. 1996. *Fundamentals of Atmospheric Physics*. San Diego: Academic Press.

Shepherd, J. M. 2006. "Evidence of Urban-Induced Precipitation Variability in Arid Climate Regimes." *Journal of Arid Environments*, 67, 607–628.

Watson, A. I., Lopez, R. A., & Holle, R. L. 1994. "Diurnal Cloud-to-Ground Lightning Patterns in Arizona during the Southwest Monsoon." *Monthly Weather Review*, 122, 1716–1725.

7

Earth's Radiation Budget

One can make a day of any size, and regulate the rising and setting of his own sun and the brightness of its shining.

—John Muir, 1875

> The sanguine Sunrise, with his meteor eyes,
> And his burning plumes outspread,
> Leaps to the back of my sailing rack,
> When the morning star shines dead;
> As on the jag of a mountain crag,
> Which an earthquake rocks and swings,
> An eagle alit one moment may sit
> In the light of the golden wings.
> And when Sunset may breathe, from the lit sea beneath,
> Its ardours of rest and of love,
> And the crimson pail of eve may fall
> From the depth of Heaven above,
> With wings folded I rest, on mine aery nest,
> As still as a brooding dove.

—Percy Bysshe Shelley, *The Cloud*, 1820

BEFORE BEGINNING A BASIC DISCUSSION of energy transfer and earth's energy budget it is important to have a fundamental knowledge and understanding of energy and thermal heat principles. In the following sections, a brief introduction to the fundamentals is provided.

Energy

(Note: Material in this section is from F. R. Spellman, 2008, *Physics for the Non-Physicists.* Boca Raton, FL: CRC Press.) Energy (often defined as the ability to do work) is one of the most discussed topics today because of high prices for hydrocarbon products (gasoline and diesel fuel), electricity, and natural gas. These are all forms of energy that we are quite familiar with, but energy also comes in other forms—heat (thermal), light (radiant), mechanical, and nuclear energy. Energy is in everything. All things we do in life and death (biodegradation requires energy, too) are a result of energy. But remember that the ultimate source of energy on earth is obtained from the sun. There are two types of energy—stored (potential) energy and working or moving (kinetic) energy.

Potential Energy

An object can have the ability to do work—to have energy—because of position. For example, a weight suspended high from a scaffold can be made to exert a force when it falls. Because gravity is the ultimate source of this energy it is correctly called gravitational potential energy or GPE (GPE = weight × height), but we usually refer to this as potential energy or PE.

Another type of potential energy is chemical potential energy—the energy stored in a battery or the gas in a vehicle's gas tank.

Consider Figure 7.1. When the suspended object is released, it will fall on top of the box and crush or squash it, exerting a force on the box over a distance. By multiplying the force exerted on the box by the distance the object falls, we could calculate the amount of work that is done.

Kinetic Energy

Moving objects have energy (ability to do work). Kinetic energy of an object is related to its motion. Figure 7.2 shows the suspended box of bricks we used earlier to demonstrate potential energy but now the box of bricks is free-falling—the potential energy is converted to kinetic energy because of movement. Specifically, *kinetic energy* (KE) of an object is defined as half its mass times its velocity squared, or KE = $\frac{1}{2}$ mv^2.

FIGURE 7.1
A box of bricks (gravitational potential energy, GPE) suspended
above an empty cardboard box.

FIGURE 7.2
Free-falling box of bricks (kinetic energy, KE) suspended above
an empty cardboard box.

From this equation, it is apparent that the more massive an object, and the faster it is moving, the more kinetic energy it possesses. The units of kinetic energy are determined by taking the product of the units for mass (kg) and velocity squared (m^2/s^2)—the units of KE, like PE, are joules. Kinetic energy can never be negative and only tells us about speed, not velocity.

Thermal Properties

Thermal properties of chemicals and other substances are important in physics. Such knowledge is used in math calculations, in the study of the physical properties of materials, in hazardous materials spill mitigation, and in solving many other complex environmental problems. *Heat* is a form of energy—thermal energy that can be transferred between two bodies that are at different temperatures. Whenever work is performed, usually a substantial amount of heat is caused by friction. The conservation of energy law tells us the work done plus the heat energy produced must equal the original amount of energy available. That is,

$$\text{Total energy} = \text{work done} + \text{heat produced} \qquad (7.1)$$

A traditional unit for measuring heat energy is the calorie. A **calorie** (cal) is defined as the amount of heat necessary to raise 1 gram of pure liquid water by 1° Celsius at normal atmospheric pressure. In SI units,

$$1 \text{ cal} = 4.186 \text{ J (joule)}$$

The calorie we have defined should not be confused with the one used when discussing diets and nutrition. A kilocalorie is 1,000 calories as we have defined it—the amount of heat necessary to raise the temperature of 1 kilogram of water by 1°C.

In the British system of units, the unit of heat is the British thermal unit, or Btu. One *Btu* is the amount of heat required to raise 1 pound of water 1° Fahrenheit at normal atmospheric pressure (1 atm).

Heat can be transferred in three ways: by conduction, convection, and radiation. When direct contact between two physical objects at different temperatures occurs, heat is transferred via **conduction** from the hotter object to the colder one. When a gas or liquid is placed between two solid objects, heat is transferred by **convection**. Heat is also transferred when no physical medium exists by **radiation** (e.g., radiant energy from the sun). Conduction, convection, and radiation heat transfer will be discussed in detail.

Specific Heat

Earlier we pointed out that 1 kilocalorie of heat is necessary to raise the temperature of 1 kilogram of water 1° Celsius. Other substances require different amounts of heat to raise the temperature of 1 kilogram of the substance 1°. The *specific heat* of a substance is the amount of heat in kilocalories necessary to raise the temperature of 1 kilogram of the substance 1° Celsius.

The units of specific heat are Kcal/kg °C or, in SI units, J/kg °C. The specific heat of pure water, for example, is 1.000 kcal/kg °C, or 4186 J/kg °C.

The greater the specific heat of a material, the more heat is required. Also, the greater the mass of the material or the greater the temperature change desired, the more heat that is required.

If the specific heat of a substance is known, it is possible to calculate the amount of heat required to raise the temperature of that substance. In general, the amount of heat required to change the temperature of a substance is proportional to the mass of the substance and the change in temperature, according to the following relation:

$$Q = mc \, \Delta T \tag{7.2}$$

where Q = heat required, m = mass of the substance, c = specific heat of the substance, and ΔT = change in temperature.

The amount of heat necessary to change 1 kilogram of a solid into a liquid at the same temperature is called the *latent heat of fusion* of the substance. The temperature of the substance at which this change from solid to liquid takes place is known as the *melting point*. The amount of heat necessary to change 1 kilogram of a liquid into a gas is called the *latent heat of vaporization*. When this point is reached, the entire mass of substance is in the gas state. The temperature of the substance at which this change from liquid to gas occurs is known as the *boiling point.*

Transferring Energy

There are three ways of transferring energy:

Convection is dynamic flow of heat through a bulk, macroscopic movement of matter from a hot to a cool region, as opposed to the microscopic transfer of heat between atoms involved with conduction. Convection is the least efficient method of heat transfer. A good example of transfer of heat by convection is heating a pot of water on the stove. When water in the bottom

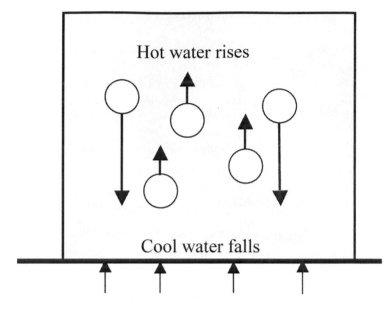

FIGURE 7.3
Convection currents in boiling water.

of the pot is heated, the heat is transferred from the hot water at the bottom to the cooler water at the top by convection. These convection currents are illustrated in Figure 7.3.

Conduction is energy flow through solids as each molecule transfers the energy to the next like a "pass it on line"; the electron pushes or transfers its energy to the electrons in line in an electrical conductor causing electric current flow.

Radiation is the most efficient way of heating. Every object that has a temperature radiates energy. Radiated energy increases are felt 1,000 times more by people than changes through means of conduction or convection.

Did You Know?

One calorie is a unit of energy that can raise 1 gram of water 1°C. One kilocalorie is the dietary calorie about which people are so concerned.

Earth's Heat Balance

The energy expended in virtually all atmospheric processes is originally derived from the sun. This energy is transferred by radiation of heat in the form of electromagnetic waves. The radiation from the sun has its peak energy transmission in the visible wavelength range [0.38 to 0.78 micrometers (μm)] of the electromagnetic spectrum. However, the sun also releases considerable energy in the ultraviolet and infrared regions. Ninety-nine percent of the sun's energy is emitted in wavelengths between 0.15 and 40 μm. Furthermore, wavelengths longer than 2.5 μm are strongly absorbed by water vapor and carbon dioxide in the atmosphere. Radiation at a wavelength less than 0.29 μm is absorbed high in the atmosphere by nitrogen and oxygen. Therefore, solar radiation striking the earth generally has a wavelength between 0.29 and 2.5 μm. It is important to keep in mind that the sun is too hot to burn. It fuses hydrogen into helium, and during this fusion process, 3% of the mass of the two atoms of hydrogen is turned into heat. This emits 56×10^{26} calories per minute, like a 6,000°K black body.

Since energy from the sun is always entering the atmosphere, the earth would overheat if all this energy were stored in the earth-atmosphere system. So, energy must eventually be released back into space. On the whole, this is what happens—approximately 50% of the solar radiation entering the atmosphere reaches earth's surface, either directly or after being scattered by clouds, particulate matter, or atmospheric gases. The other 50% is either reflected directly back or absorbed in the atmosphere and its energy reradiated back into space at a later time as infrared radiation. Most of the solar energy reaching the surface is absorbed and must be returned to space to maintain *heat balance* (aka *radiational balance*). The energy produced within the earth's interior (from the hot mantle area via convection and conduction) that reaches the earth's surface (about 1% of that received from the sun) must also be lost.

Earth's Radiation Budget

(From NASA 2007: *What Is the Earth's Radiation Budget?* Accessed July 5, 2011, at http://eosweb.larc.nasa.gov/EDDOCS/whatis.html.) The earth's radiation budget is a concept used for understanding how much energy the earth gets from the sun and how much energy the earth-system radiates back to outer space as invisible light.

If the earth and the earth's atmosphere retain more solar energy than they radiate back to space, the earth will warm. If the earth and the earth-system radiate more energy to space than receive from the sun, the earth will cool.

FIGURE 7.4
Just the right balance!

Scientists think of the radiation budget in terms of a see-saw or balance (see Figure 7.4). If the earth retains more energy from the sun, the earth warms and emits more infrared energy. This brings the earth's radiation budget into balance. If the earth emits more of this energy than it absorbs, the earth cools. As it cools, the earth emits less energy. This change also brings the radiation budget back into balance.

Absorbed sunlight raises the earth's temperature. Emitted radiation or heat lowers the temperature. When absorbed sunlight and emitted heat balance each other, the earth's temperature doesn't change—the radiation budget is in balance.

NASA 2007 points out that there are three basic parts of the radiation budget: solar incident energy, solar reflected energy, and earth emitted energy (see Figure 7.5). Incoming solar radiation is absorbed by the earth's surface, water vapor, gases, and aerosols in the atmosphere. This incoming solar radiation is also reflected by the earth's surface, by clouds, and by the atmosphere. Energy that is absorbed is emitted by the earth-atmosphere system as long-wave radiation.

Reradiation

Reradiation of energy from the earth is accomplished by three energy transport mechanisms: radiation, conduction, and convection. *Radiation* of energy, as stated earlier, occurs through electromagnetic radiation in the infrared region of the spectrum. The crucial importance of the radiation mechanism is that it carries energy away from earth on a much longer wavelength than the solar energy (sunlight) that brings energy to the earth and, in turn, works to maintain the earth's heat balance. The earth's heat balance

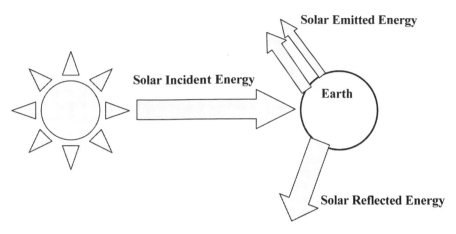

FIGURE 7.5
Earth radiation components.

is of particular interest to us in this text because it is susceptible to upset by human activities.

A comparatively smaller but significant amount of heat energy is transferred to the atmosphere by conduction from the earth's surface. *Conduction* of energy occurs through the interaction of adjacent molecules with no visible motion accompanying the transfer of heat; for example, the whole length of a metal rod will become hot when one end is held in a fire. Because air is a poor heat conductor, conduction is restricted to the layer of air in direct contact with the earth's surface. The heated air is then transferred aloft by *convection*, the movement of whole masses of air, which may be either relatively warm or cold. Convection is the mechanism by which abrupt temperature variations occur when large masses of air move across an area. Air temperature tends to be greater near the surface of the earth and decreases gradually with altitude. A large amount of the earth's surface heat is transported to clouds in the atmosphere by conduction and convection before being lost ultimately by radiation, and this redistribution of heat energy plays an important role in weather and climate conditions.

The earth's average surface temperature is maintained at about 15°C because of atmospheric greenhouse effect. Greenhouse effect occurs when the gases of the lower atmosphere transmit most of the visible portion of incident sunlight in the same way as the glass of a garden greenhouse. The warmed earth emits radiation in the infrared region, which is selectively absorbed by the atmospheric gases whose absorption spectrum is similar to that of glass. This absorbed energy heats the atmosphere and helps maintain the earth's temperature. Without this greenhouse effect, the surface temperature would average around –18°C. Most of the absorption of infrared energy is performed by water molecules in the atmosphere. In addition to the key role played by water molecules, carbon dioxide, although to a lesser extent, also is essential in maintaining the heat balance. Environmentalists and others concerned with environmental issues are concerned that an increase in the carbon dioxide level in the atmosphere could prevent sufficient energy loss, causing damaging increases in the earth's temperature. This phenomenon, commonly known as anthropogenic greenhouse effect, may occur from elevated carbon dioxide levels caused by increased use of fossil fuels and the reduction in carbon dioxide absorption because of destruction of the rainforest and other forest areas.

Insolation (EPA, 2005)

The amount of incoming solar radiation received at a particular time and location in the earth-atmosphere system is called insolation. Insolation is governed by four factors:

1. Solar constant
2. Transparency of the atmosphere
3. Daily sunlight duration
4. Angle at which the sun's rays strike the earth

Solar Constant

The *solar constant* is the average amount of radiation received at a point, perpendicular to the sun's rays, that is located outside the earth's atmosphere at the earth's mean distance from the sun. The average amount of solar radiation received at the outer edge of the atmosphere would vary slightly depending on the energy output of the sun and the distance of the earth relative to the sun. Due to the eccentricity of the earth's orbit around the sun, the earth is closer to the sun in January than in July. Also, the radiation emitted from the sun varies slightly, probably less than a few percent. These slight variations that affect the solar constant are trivial considering the atmospheric properties that deplete the overall amount of solar radiation reaching the earth's surface. Transparency of the atmosphere, duration of daylight, and the angle at which the sun's rays strike the earth are much more important in influencing the amount of radiation actually received, which in turn includes the weather.

Transparency

Transparency of the atmosphere does have an important bearing upon the amount of insolation that reaches the earth's surface. The emitted radiation is depleted as it passes through the atmosphere. Different atmospheric constituents absorb or reflect energy in different ways and in varying amounts. Transparency of the atmosphere refers to how much radiation penetrates the atmosphere and reaches the earth's surface without being depleted.

Did You Know?

Some of the radiation received by the atmosphere is reflected from the tops of clouds and from the earth's surface and some is absorbed by molecules and clouds.

The general reflectivity of the various surfaces of the earth is referred to as the albedo. Albedo is defined as the fraction (or percentage) of incoming solar energy that is reflected back to space. Different surfaces (water, snow, sand, etc.) have different albedo values as discussed later. For the earth and atmosphere as a whole, the average albedo is 30% for average conditions of cloudiness over the earth. This reflectivity is greatest in the visible range of wavelengths.

Some of the gases in the atmosphere (notably water vapor) absorb solar radiation, causing less radiation to reach the earth's surface. Water vapor, although comprising only about 3% of the atmosphere, on the average absorbs about six times as much solar radiation as all other gases combined. The amount of radiation received at the earth's surface is therefore considerably less than that received outside the atmosphere as represented by the solar constant.

Did You Know?

The earth warms up when it absorbs energy and cools when it radiates energy. The earth absorbs and emits radiation at the same time. If the earth's surface absorbs more energy than it radiates, it will heat up. If the earth's surface radiates more energy than it absorbs, it will cool.

Daylight Duration

The duration of daylight also affects the amount of insolation received: The longer the period of sunlight, the greater the total possible insolation.

Daylight duration varies with latitude and the seasons. At the equator, day and night are always equal. In the polar regions, the daylight period reaches a maximum of 24 hours in summer and a minimum of zero hours in winter.

Angle of Sun's Rays

The angle at which the sun's rays strike the earth varies considerably as the sun "shifts" back and forth across the equator. A relatively flat surface perpendicular to an incoming vertical sun ray receives the largest amount of insolation. Therefore, areas at which the sun's rays are oblique receive less insolation because the oblique rays must pass through a thicker layer of reflecting and absorbing atmosphere and are spread over a greater surface area. This same principle also applies to the daily shift of the sun's rays. At solar noon, the intensity of insolation is greatest. In the morning and evening hours, when the sun is at a low angle, the amount of insolation is small.

Heat Distribution

The earth, as a whole, experiences great contrasts in heat and cold at any particular time. Warm, tropical breezes blow at the equator while ice caps are forming in the polar regions. In fact, due to the extreme temperature differences at the equator and the poles, the earth-atmosphere system resembles a giant "heat engine." Heat engines depend on hot-cold contrasts to generate power. As you will see, this global "heat engine" influences the major atmospheric circulation patterns as warm air is transferred to cooler areas. Different parts of the earth receiving different amounts of insolation account for much of this heat imbalance. As discussed earlier, latitude, the seasons, and daylight duration cause different locations to receive varying amounts of insolation.

Differential Heating

Not only do different amounts of solar radiation reach the earth's surface, but different earth surfaces absorb heat energy at different rates. For example, land masses absorb and store heat differently from water masses. Also, different types of land surfaces vary in their ability to absorb and store heat. The color, shape, surface texture, vegetation, and presence of buildings can all influence the heating and cooling of the ground. Generally, dry surfaces heat and cool faster than moist surfaces. Plowed fields, sandy beaches, and paved roads become hotter than surrounding meadows and wooded areas. During the day, the air over a plowed field is warmer than over a forest or swamp; during the night, the situation is reversed. The property of different surfaces that causes them to heat and cool at different rates is referred to as *differential heating.*

Absorption of heat energy from the sun is confined to a shallow layer of land surface. Consequently, land surfaces heat rapidly during the day and cool quickly at night. Water surfaces, on the other hand, heat and cool more slowly than land surfaces for the following reasons:

- Water movement distributes heat.
- The sun's rays are able to penetrate the water surface.
- More heat is required to change the temperature of water due to its higher specific heat. (It takes more energy to raise the temperature of water than it does to change the temperature of the same amount of soil.)
- Evaporation of water occurs, which is a cooling process.

Transport of Heat

Earlier, it was pointed out that in addition to radiation, heat is transferred by conduction, convection, and advection. These processes affect the temperature of the atmosphere near the surface of the earth. *Conduction* is the process by which heat is transferred through matter without the transfer of matter itself. For example, the handle of an iron skillet becomes hot due to the conduction of heat from the stove burner. Heat is conducted from a warmer object to a cooler one. Heat transfer occurs when matter is in motion. Air that is warmed by a heated land surface (by conduction) will rise because it is lighter than the surrounding air. This heated air rises, transferring heat vertically. Likewise, cooler air aloft will sink because it is heavier than the surrounding air. This goes hand in hand with rising air and is part of heat transfer by convection. Meteorologists also use the term *advection* to denote heat transfer that occurs mainly by horizontal motion rather than by vertical movement or air (convection).

Global Distribution of Heat

As mentioned before, the world distribution of insolation is closely related to latitude. Total annual insolation is greatest at the equator and decreases toward the poles. The amount of insolation received annually at the equator is over four times that received at either of the poles. As the rays of the sun shift seasonally from one hemisphere to the other, the zone of maximum possible daily insolation moves with them. For the earth as a whole, the gains in solar energy equal the losses of energy back into space (heat balance). However, since the equatorial region does gain more heat than it loses and the poles lose more heat than they gain, something must happen to distribute heat more evenly around the earth. Otherwise, the equatorial regions would continue to heat and the poles would continue to cool. Therefore, in order to reach equilibrium, a continuous large-scale transfer of heat (from low to high altitudes) is carried out by atmospheric and oceanic circulations.

The atmosphere drives warm air poleward and brings cold air toward the equator. Heat transfer from the tropics poleward takes place throughout the year, but at a much slower rate in summer than in winter. The temperature difference between low and high latitudes is considerably smaller in summer than in winter (only about half as large in the northern hemisphere). As would be expected, the winter hemisphere has a net energy loss and the summer hemisphere a net gain. Most of the summertime gain is stored in the surface layers of land and ocean, but mainly in the ocean.

The oceans also play a role in heat exchange. Warm water flows poleward along the western side of an ocean basin and cold water flows toward the equator on the eastern side. At higher latitudes, warm water moves poleward in the eastern side of the ocean basin and cold water flows toward the equator on the western side. The oceanic currents are responsible for about 40% of the transport of energy from the equator to the poles. The remaining 60% is attributed to the movement of air.

Did You Know?

A dyne is a unit of force describing the condition equal to 1 gram of mass accelerated 1 centimeter per second. Force per unit area is pressure. The atmosphere generates the force or pressure of 1,013,250 dynes/cm^2. A bar is a pressure measurement of the magnitude of 10^6 and a millibar is 1/1,000 of a bar. Sea level pressure, millibars, is 1,013.25.

Albedo

As already mentioned, *albedo* is the ratio between the light reflected from a surface and the total light falling on it. Albedo is a surface phenomenon— basically a radiation reflector. Albedo always has a value less than or equal

TABLE 7.1
The Albedo of Some Surface Types in % Reflected

Surface	Albedo
Water (low sun)	10–100
Water (high sun)	3–10
Grass	16–26
Glacier ice	20–40
Deciduous forest	15–20
Coniferous forest	5–15
Old snow	40–70
Fresh snow	75–95
Sea ice	30–40
Blacktopped tarmac	5–10
Desert	25–30
Crops	15–25

to 1. An object with a high albedo, near 1, is very bright, while a body with a low albedo, near 0, is dark. For example, freshly fallen snow typically has an albedo that is between 75% and 90%; that is, 75% to 95% of the solar radiation that is incident on snow is reflected. At the other extreme, the albedo of a rough, dark surface, such as a green forest, may be as low as 5%. The albedos of some common surfaces are listed in Table 7.1. The portion of insolation not reflected is absorbed by the earth's surface, warming it. This means earth's albedo plays an important part in the earth's radiation balance and influences the mean annual temperature and the climate on both local and global scales.

Scattering

In addition to the radiation reflecting properties of albedo, another phenomenon known as scattering occurs in the atmosphere. Small particles scatter light differently than large particles. A particle one-tenth of a micron in diameter, for instance, will scatter light selectively. Note that $1 \, \mu = 10^{-6}$ meter = .000001 meter = 1 micron = 1. In the atmosphere, the scattering refers to the various colors of the electromagnetic spectrum, consisting of the colors red, orange, yellow, green, blue, indigo, and violet or ROY G BIV. Because violet is the first color scattered in the upper atmosphere, then indigo, then blue, and so forth, the spectrum order of color from shortest wavelengths (λ) to longer wavelengths is listed as VIB G YOR.

Did You Know?

When the shorter wavelengths of light are removed, we can see the sun and objects by means of those longer wavelengths—we see the yellow-orange-red part of the spectrum. Apparently, the short wavelengths of light can't get around the particles in the atmosphere but the longer ones can.

Absorption

All short wavelengths (λ's) are absorbed in the thermosphere and all the middle ultraviolet wavelengths are absorbed in the stratosphere, protecting us from dangerous radiation. Moreover, in addition to protecting us from dangerous UV radiation, the atmosphere is also a stabilizer of temperature. Note that the albedo of the moon is only about 7%; it is a good absorber, but the moon's temperature is about +250°F during the daytime and –250°F at night. In contrast, the earth's albedo is about 34% and there is never more than a 90°F difference for any location between night and day.

Did You Know?

Water vapor and carbon dioxide in the atmosphere are two gases that are excellent absorbers of UV radiation. Moreover, the atmosphere absorbs about 18% of the sun's incident energy—2% of the short wavelengths in the upper atmosphere and 16% of the middle and long wavelengths in the stratosphere and troposphere. Also, between 34% and 35% of the sun's energy is reflected. Doing the math: $18 + 35 = 53\%$ of the sun's energy never gets down to the earth's surface.

Climate Change and Greenhouse Effect

(This section is based on information in F. R. Spellman, 2009, *The Science of Environmental Pollution,* 2nd ed., Boca Raton, FL. CRC Press.) Is earth's climate changing? Are warmer times or colder times on the way? Is the greenhouse effect going to affect our climate, and if so, do we need to worry about it? Will the tides rise and flood New York? Does the ozone hole portend disaster right around the corner?

These and many other questions related to climate change have come to the attention of us all. A constant barrage of newspaper headlines, magazine articles, best-selling so-called "science-based convenient truth books," and television news reports on these topics have inundated us, have warned us, and have scared the bejesus out of many of us. Recently, we've seen constant reports on El Niño and its devastation of the west coast of the United States and Peru and Ecuador—and its reduction of the number, magnitude, and devastation of hurricanes that annually blast the east coast of the United States.

Scientists have been warning us of the catastrophic harm that can be done to the world by atmospheric warming. One view states that the effect could bring record droughts, record heat waves, record smog levels, and an increasing number of forest fires.

Another caution put forward warns that the increasing atmospheric heat could melt the world's ice caps and glaciers, causing ocean levels to rise to the point where some low-lying island countries would disappear, while the coastlines of other nations would be drastically altered for ages—or perhaps for all time.

In regard to global climate change, what's going on? We hear plenty of theories put forward by doomsayers, but are they correct? If they are correct, what does it all mean? Does anyone really know the answers? Should we be concerned? Should we expect inconvenience on a global scale? Should we invest in waterfront property in Antarctica? Should we panic? Should we take the space shuttles out of mothballs and head for other planets?

No. While no one really knows the answers, and while we should be concerned, no real cause for panic exists. Every problem has a solution—that is, when cool heads and science prevail. So, the questions are these: Should we take some type of decisive action, any type of action? Should we come up with quick answers and put together a plan to fix these problems? What really needs to be done? What can we do? Is there anything we can do?

The key question to answer here is, What really needs to be done? We can study the facts, the issues, the possible consequences—but the key to successfully combating these issues is to stop and seriously evaluate the problems. Again, we need to let scientific fact, common sense, and cool-headedness prevail. Shooting from the hip is not called for and makes little sense—and could have Titanic consequences for us all. Remember, we are all riding one ship through life and that ship is earth.

The other question that has merit here is, Will we take the correct actions before it is too late? The key words here are "correct actions." Eventually, we will have to take action. But we do not yet know what those actions should be.

In this section, we discuss global climate change related to our atmosphere and its problems, actual and potential. Consider this: Any damage we do to our atmosphere affects the other three environmental mediums: water, soil, and biota (us). Thus, the endangered atmosphere (if it is endangered) is a major concern to all of us. When it comes to a definitive answer on whether we are experiencing human-caused global climate change and what that

TABLE 7.2
Geologic Eras and Periods

Era	Period	Millions of Years before Present
Cenozoic	Quaternary	2.5–present
	Tertiary	65–2.5
Mesozoic	Cretaceous	135–65
	Jurassic	190–135
	Triassic	225–190
Paleozoic	Permian	280–225
	Pennsylvanian	320–280
	Mississippian	345–320
	Devonian	400–345
	Silurian	440–400
	Ordovician	500–440
	Cambrian	570–500
Precambrian (supereon)		4,600–570

could mean to life on earth, we simply do not know what we do not know about global climate change. In this text, it is the convenient truth (based on real science, not feel-good science, and real research) and not some wacko theory we are concerned with.

The Past

Before we begin our discussion of the past, we need to define the era we refer to when we say "the past." Tables 7.2 and 7.3 are provided to assist us in making this definition. Table 7.2 gives the entire expanse of time from earth's beginning to present. Table 7.2 provides the sequence of geological epochs

TABLE 7.3
Epochs

Epoch	Million Years Ago
Holocene	0.01
Pleistocene	1.6
Pliocene	5
Miocene	24
Oligocene	35
Eocene	58
Paleocene	65

over the past 65 million years, as dated by modern methods. The Paleocene through Pliocene together make up the Tertiary period; the Pleistocene and the Holocene compose the Quaternary period.

When we think about climatic conditions in the prehistoric past, two things generally come to mind—ice ages and dinosaurs. Of course, in the immense span of time prehistory covers, those two eras represent only a brief moment in time, so let's look at what we know about the past and about earth's climate and conditions. One thing to consider—geological history shows us that in the past the "normal" climate of the earth was so warm that subtropical weather reached to 60°N and S, and polar ice was entirely absent.

Only during less than about 1% of the earth's history did glaciers advance and reach as far south as what is now the temperate zone of the northern hemisphere. The latest such advance, which began about 1 million years ago, was marked by geological upheaval and (perhaps) the advent of human life on earth. During this time, vast ice sheets advanced and retreated, grinding their way over the continents.

A Time of Ice

Nearly 2 billion years ago, the oldest known glacial epoch occurred. A series of deposits of glacial origin in southern Canada, extending east to west about 1,000 miles, shows us that within the last billion years or so, apparently at least six major phases of massive, significant climatic cooling and consequent glaciation occurred at intervals of about 150 million years. Each lasted perhaps as long as 50 million years.

Examination of land and oceanic sediment core samples clearly indicate that in more recent times (the Pleistocene epoch to the present), many alternating episodes of warmer and colder conditions occurred over the last 2 million years (during the middle and early Pleistocene epochs). In the last million years, at least eight such cycles have occurred, with the warm part of the cycle lasting a relatively short interval.

During the Great Ice Age (the Pleistocene epoch), ice advances began, a series of them that at times covered over one-quarter of the earth's land surface. Great sheets of ice thousands of feet thick, glaciers, moved across North America over and over, reaching as far south as the Great Lakes. An ice sheet thousands of feet thick spread over Northern Europe, sculpting the land and leaving behind lakes, swamps, and terminal moraines as far south as Switzerland. Each succeeding glacial advance was apparently more severe than the previous one. Evidence indicates that the most severe began about 50,000 years ago and ended about 10,000 years ago. Several interglacial stages separated the glacial advances, melting the ice. Average temperatures were higher than ours today.

"Temperatures were higher than today?" Yes. *Think about that* as we proceed.

Because one-tenth of the globe's surface is still covered by glacial ice, scientists consider the earth still to be in a glacial stage. The ice sheet has been retreating since the climax of the last glacial advance and world climates, although fluctuating, are slowly warming.

From our observations and from well-kept records, we know that the ice sheet is in a retreating stage. The records clearly show that a marked worldwide retreat of ice has occurred over the last hundred years. World famous for its 50 glaciers and 200 lakes, Glacier National Park in Montana does not present the same visual experiences it did a hundred years ago. A 10-foot pole put into place at the terminal edge of one of the main glaciers holds a "1939" sign. The sign is still in place—but the terminal end of the glacier has retreated several hundred feet back up the slope of the mountain. Swiss resorts built during the early 1900s to offer scenic glacial views now have no ice in sight. Theoretically, if glacial retreat continues, melting all of the world's ice supply, sea levels would rise more than 200 feet, flooding many of the world's major cities. New York and Boston would become aquariums.

The question of what causes ice ages is one scientists still grapple with. Theories range from changing ocean currents to sunspot cycles. Of one fact we are absolutely certain, however. An ice age event occurs because of a change in earth's climate. But what could cause such a drastic change?

Climate results from uneven heat distribution over earth's surface. It's caused by the earth's tilt—the angle between the earth's orbital plane around the sun and its rotational axis. This angle is currently 23.5°.

The angle has changed. It has not always been 23.5°. As pointed out earlier, the angle, of course, affects the amount of solar energy that reaches the earth and where it falls.

The heat balance of the earth, which is driven mostly by the concentration of carbon dioxide (CO_2) in the atmosphere, also affects long-term climate.

If the pattern of solar radiation changes and/or if the amount of CO_2 changes, climate change can result. Abundant evidence that the earth does undergo climatic change exists, and we know that climatic change can be a limiting factor for the evolution of many species.

Evidence (primarily from soil core samples and topographical formations) tells us that change in climate includes events such as periodic ice ages characterized by glacial and interglacial periods. Long glacial periods lasted up to 100,000 years, where temperatures decreased about 9°F and ice covered most of the planet. Short periods lasted up to 12,000 years, with temperatures decreasing by 5°F, and ice covering 40° latitude and above. Smaller periods (the "Little Ice Age," which occurred from about 1000–1850 AD) had about a 3°F drop in temperature. (Note: Despite its name, "Little Ice Age" was a time of severe winters and violent storms, not a true glacial period.)

These ages may be or not be significant—but consider that we are presently in an interglacial period and that we may be reaching its apogee. What does that mean? No one knows with any certainty. Let's look at the effects of ice ages—the effects we think we know about.

Changes in sea levels could occur. Sea level could drop by about 100 meters in a full-blown ice age, exposing the continental shelves. Increased deposition during melt would change the composition of the exposed continental shelves. Less evaporation would change the hydrological cycle. Significant landscape changes could occur—on the scale of the Great Lakes formation. Drainage patterns throughout most of the world and topsoil characteristics would change. Flooding on a massive scale could occur.

How would these changes affect us? That depends—on whether you live in Northern Europe, Canada, Seattle, or Washington; around the Great Lakes; or near a seashore.

We are not sure what causes ice ages, but we have some theories (don't people always have theories?). To generate a full-blown (massive ice sheet covering most of the globe) ice age, scientists point out that certain periodic or cyclic events or happenings must occur. Periodic fluctuations would have to affect the solar cycle, for instance. However, we have no definitive proof that this has ever occurred.

Another theory speculates periods of volcanic activity could generate masses of volcanic dust that would block or filter heat from the sun. This would cool down the earth. Some speculate that the carbon dioxide cycle would have to be periodic/cyclic to bring about periods of climate change. There is reference to a so-called factor 2 reduction, causing a 7°F temperature drop worldwide. Others speculate that another global ice age could be brought about by increased precipitation at the poles due to changing orientation of continental land masses. Others theorize that a global ice age would result if changes in the mean temperatures of ocean currents occurred. But the question is how? By what mechanism? Are these plausible theories? No one is sure—this is speculation.

Speculation aside, what are the most probable causes of ice ages on earth? According to the Milankovitch hypothesis, ice age occurrences are governed by a combination of factors: (1) the earth's change of altitude in relation to the sun (the way it tilts in a 41,000-year cycle and at the same time wobbles on its axis in a 22,000-year cycle) making the time of its closest approach to the sun come at different seasons; and (2) the 92,000-year cycle of eccentricity in its orbit round the sun, changing it from an elliptical to a near circular orbit, the severest period of an ice age coinciding with the approach to circularity.

So what does all this mean? We have a lot of speculation about ice ages and their causes and their effects. This is the bottom line. We know that ice ages occurred—we know that they caused certain things to occur (formation of the Great Lakes, etc.), and while there is a lot we do not know, we recognize the possibility of recurrent ice ages.

Lots of possibilities exist. Right now, no single theory is sound, and doubtless many factors are involved. Keep in mind that the possibility does exist that we are still in the Pleistocene ice age. It may reach another maximum in another 60,000 plus years or so.

Warm Winter

When we discussed possible causes of glaciation and subsequent climatic cooling, we were left hanging on the standard syndrome: We don't know what we don't know. In this section, we discuss how we know what we think we do know about climatic change.

The headlines we see in the paper sound authoritative: "1997 Was the Warmest Year on Record." "Scientists Discover Ozone Hole Is Larger Than Ever." "Record Quantities of Carbon Dioxide Detected in Atmosphere." Or maybe you saw the one that read "January 1998 Was the Third Warmest January on Record." Other reports indicate we are undergoing a warming trend. But conflicting reports abound. What do we know about climate change?

What We Think We Know about Global Climate Change

Two environmentally significant events took place late in 1997: El Niño's return and the Kyoto Conference: Summit on Global Warming and Climate Change. News reports blamed El Niño for just about anything that had to do with weather conditions throughout the world. Some occurrences were indeed El Niño–related or generated: the out-of-control fires, droughts, and floods; the stretches of dead coral, no sign of fish in the water, and few birds around certain Pacific atolls. The devastating storms that struck the west coasts of South America, Mexico, and California were also probably El Niño related. El Niño's effect on the 1997 hurricane season, one of the mildest on record, is not in question, either.

But does a connection exist between El Niño and global warming or global climate change? On December 7, 1997, the Associated Press reported that while delegates at the global climate conference in Kyoto haggle over greenhouse gases and emission limits, a compelling question has emerged: "Is global warming fueling El Niño?" Nobody knows for sure.

Why aren't we sure? Because we need more information than we have today. Our paltry amount of recorded data, however, suggest that El Niño is getting stronger and more frequent.

Some scientists fear that El Niño's increasing frequency and intensity (records show that two of this century's three worst El Niños came in 1982 and 1997) may be linked to global warming. At the Kyoto Conference, experts said the hotter atmosphere is heating up the world's oceans, setting the stage for more frequent and extreme El Niños.

Weather-related phenomena seem to be intensifying throughout the globe. Can we be sure that this is related to global warming yet? No. Without more data, more time, and more science, we cannot be sure.

Should we be concerned? Yes. According to the Associated Press coverage of the Kyoto Conference, scientist Richard Fairbanks reported that he found startling evidence of our need for concern. During two months of scientific experiments on Christmas Island (the world's largest atoll in the Pacific Ocean) conducted in autumn 1997, he discovered a frightening scene. The water surrounding the atoll was 7°F higher than average for the time of year, which upset the balance of the environmental system. According to Fairbanks, 40% of the coral was dead, the warmer water had killed off or driven away fish, and the atoll's plentiful bird population was almost completely gone.

El Niños have an acute impact on the globe; that is not in doubt. However, we do not know if they are caused by or intensified because of global warming. What do we know about global warming and climate change?

USA Today (December, 1997) discussed the results of a report issued by the Intergovernmental Panel on Climate Change. They interviewed Jerry Mahlman, of the National Oceanic and Atmospheric Administration and Princeton University, and presented the following information about what most scientists agree on:

- There is a natural "greenhouse effect" and scientists know how it works—and without it, earth would freeze.
- The earth undergoes normal cycles of warming and cooling on grand scales. Ice ages occur every 20,000 to 100,000 years.
- Globally, average temperatures have risen 1°F in the past 100 years, within the range that might occur normally.
- The level of man-made carbon dioxide in the atmosphere has risen 30% since the beginning of the Industrial Revolution in the 19th century, and it is still rising.
- Levels of man-made carbon dioxide will double in the atmosphere over the next 100 years, generating a rise in global average temperatures of

about 3.5°F (larger than the natural swings in temperature that have occurred over the past 10,000 years).
- By 2050, temperatures will rise much higher in northern latitudes than
  the increase in global average temperatures. Substantial amounts of
  northern sea ice will melt, and snow and rain in the northern hemisphere
  will increase.
- As the climate warms, the rate of evaporation will rise, further increasing warming. Water vapor also reflects heat back to Earth (*USA Today*,
  p. A-2).

What We Think We Know about Global Warming

What is global warming? To answer this question we need to discuss
"greenhouse effect." Water vapor, carbon dioxide, and other atmospheric
gases (greenhouse gases) help warm the earth. Earth's average temperature
would be closer to zero than its actual 60°F, without the greenhouse effect.
But the average temperature could increase, changing orbital climate, as gases
are added to the atmosphere. How does greenhouse effect actually work? Let's
take a closer look at this phenomenon.

Greenhouse Effect

Earth's greenhouse effect, of course, took its name because of the similarity of
effect in a standard greenhouse for plants. Because a plant greenhouse's glass
walls and ceilings are largely transparent to short-wave radiation from the
sun, surfaces and objects inside the greenhouse absorb the radiation. The radiation, once absorbed, transforms into long-wave (infrared) radiation (heat)
and radiates back from the greenhouse interior. But the glass prevents the
long-wave radiation from escaping again, absorbing the warm rays. The interior of the greenhouse becomes much warmer than the air outside because
of the heat trapped inside.

Earth and its atmosphere undergo a process very similar to this—the atmosphere acts like this to our earth, absorbing then reradiating back. Short-wave
and visible radiation reaching earth are absorbed by the surface as heat. The
long heat waves radiate back out toward space, but the atmosphere absorbs
many of them, trapping them. This natural and balanced process is essential to supporting our life systems on earth. Changes in the atmosphere can
radically change the amount of absorption (therefore the amount of heat) the
earth's atmosphere retains. In recent decades, scientists speculate that various air pollutants have caused the atmosphere to absorb more heat. At the

local level with air pollution, the greenhouse effect causes heat islands in and around urban centers, a widely recognized phenomenon.

The main contributors to this effect are the greenhouse gases: water vapor, carbon dioxide, carbon monoxide, methane, volatile organic compounds (VOCs), nitrogen oxides, chlorofluorocarbons (CFCs), and surface ozone. These gases cause a general climatic warming by delaying the escape of infrared radiation from the earth into space. Scientists stress this is a natural process—indeed, if the "normal" greenhouse effect did not exist, the earth would be 33°C cooler than it presently is (Hansen et al., 1986).

Human activities are now rapidly intensifying the natural phenomenon of earth's greenhouse effect, which may lead to problems of warming on a global scale. Much debate, confusion, and speculation about this potential consequence are under way because scientists cannot yet agree about whether the recently perceived worldwide warming trend is because of greenhouse gases, due to some other cause, or simply a wider variation in the normal heating and cooling trends they have been studying. Unchecked, the greenhouse effect may lead to significant global warming, with profound effects upon our lives and our environment. Human impact on greenhouse effect is real; it has been measured and detected. The rate at which the greenhouse effect is intensifying is now more than five times what it was during the last century (Hansen & Lebedeff, 1989).

The Bottom Line on Global Warming

Supporters of the global warming theory base their assumptions on man's altering of the earth's normal and necessary greenhouse effect. The human activities they blame for increases of greenhouse gases include burning of fossil fuels, deforestation, and use of certain aerosols and refrigerants. These gases have increased how much heat remains trapped in the earth's atmosphere, gradually increasing the temperature of the whole globe.

From information based on recent or short-term observation, many scientists note that the last decade has been the warmest since temperature recordings began in the late 19th century. They see that the general rise in temperature over the last century coincides with the Industrial Revolution and its accompanying increase in fossil fuel use. Other evidence supports the global warming theory. In places that are synonymous with ice and snow—the Arctic and Antarctica, for example—we see evidence of receding ice and snow cover.

Trying to pin down definitively whether or not changing our anthropogenic activities could have any significant effect on lessening global warming, though, is difficult. Scientists look at temperature variations over thousands and even millions of years, taking a long-term view at earth's climate. The variations in earth's climate are wide enough that they cannot definitively show that global warming is anything more than another short-term variation. Historical records have shown the earth's temperature does vary widely, growing colder with ice ages and then warming again, and because we cannot be certain of the causes of those climate changes, we cannot be certain of what appears to be the current warming trend.

Still, debate abounds for the argument that our climate is warming and our activities are part of the equation. The 1980s saw 9 of the 12 warmest temperatures ever recorded, and the earth's average surface temperature has risen approximately 0.6°C (1°F) in the last century (USEPA, 1995). *Time* magazine (1998) reports that scientists are increasingly convinced that because of the buildup in the atmosphere of carbon dioxide and other gases produced in large part by the burning of fossil fuels, the earth is getting hotter. Each month from January through July 1998, for example, set a new average global temperature record, and if that trend continued, the surface temperature of the earth could rise by about 1.8°F to 6.3°F by 2100. At the same time, others offer as evidence that the 1980s also saw three of the coldest years: 1984, 1985, and 1986.

What is really going on? We cannot be certain. Assuming that we are indeed seeing long-term global warming, we must determine what causes it. But again, we face the problem that scientists cannot be sure of the greenhouse effect's precise causes. Our current, possible trend in global warming may simply be part of a much longer trend of warming since the last ice age. We have learned much in the past two centuries of science, but little is actually known about the causes of the worldwide global cooling and warming that sent the earth through major and minor ice ages. The data we need reach back over millennia. We simply do not possess enough long-term data to support our theories.

Currently, scientists can point to six factors they think could be involved in long-term global warming and cooling:

1. Long-term global warming and cooling could result if changes in the earth's position relative to the sun occur (i.e., the earth's orbit around the sun), with higher temperatures when the two are closer together and lower when further apart.
2. Long-term global warming and cooling could result if major catastrophes (meteor impacts or massive volcanic eruptions) occur that throw pollutants into the atmosphere that can block out solar radiation.
3. Long-term global warming and cooling could result if changes in albedo (reflectivity of earth's surface) occur. If the earth's surface were more re-

flective, for example, the amount of solar radiation radiated back toward space instead of being absorbed would increase, lowering temperatures on earth.

4. Long-term global warming and cooling could result if the amount of radiation emitted by the sun changes.
5. Long-term global warming and cooling could result if the shape and relationship of the land and oceans change.
6. Long-term global warming and cooling could result if the composition of the atmosphere changes.

"If the composition of the atmosphere changes"—this final factor, of course, defines our present concern: Have human activities had a cumulative impact large enough to affect the total temperature and climate of earth? Right now, we cannot be sure. The problem concerns us, and we are alert to it, but we are not certain.

So What Is the Verdict?

We can expect winters to be longer, if global warming is occurring, and summers hotter. Over the next 100 years, sea level will rise as much as a foot or so. Is this bad? Depends upon where you live—however, keep in mind that not only could sea level rise 1 foot over the next 100 years, it could continue to do so for many hundreds of years.

Another point to consider is that we have routine global temperature measurements for only about 100 years. Even these are unreliable because instruments and methods of observation changed over that course of time.

The only conclusion we can safely draw about climate and climate change is that we do not know if drastic changes are occurring. We could be at the end of a geological ice age. Evidence indicates that during interglacials, a period of temperature increase occurs before they plunge. Are we ascending the peak temperature range? We have no way to tell. To what extent does our human activity impact climate? Have anthropogenic effects become so marked that we have affected the natural cycle of ice ages (which lasted for roughly the last 5 million years)? Maybe we just have a breathing spell of a few centuries before the next advance of the glaciers.

When news media personnel, would-be presidential candidates, and general doomsayers make their dire warnings about global climate change (specifically that the earth is getting warmer), keep in mind that (1) the transition from ice age to interglacial (a warm period) is well documented with considerable actual evidence to back it up; and (2) that in the ancient past when earth was experiencing the transition from ice age to interglacial and warming back to cooling, there were no humans (or few at best) to contribute to the normal warming and cooling cycles. If we had actually lived in the ancient past to witness this warming trend, we certainly would not have been able to put the blame for the warming trend on man. Simply put, the fact is, at that time, man made little contribution of carbon dioxide, CFCs, or any other chemical substance to earth's atmosphere.

The real bottom line: When the doomsayers spout out their fire and brimstone, their warnings of cataclysmic doom-and-gloom scenarios, or try to sell you waterfront property in Antarctica, remember one salient point: Would you rather experience global warming that we can adjust to with some effort or would you rather experience global cooling (another ice age) when we have the energy problems we have today? Simply, where would we find the fuel to keep warm?

References and Recommended Reading

Associated Press. 1997. "Does Warming Feed El Niño?" *Virginian-Pilot* (Norfolk, VA), p. A-15, December.

Associated Press. 1998. "Ozone Hole Over Antarctica at Record Size." *Lancaster New Era* (Lancaster, PA), September 28.

Associated Press. 1998. "Tougher Air Pollution Standards Too Costly, Midwestern States Say." *Lancaster New Era* (Lancaster, PA), September 25.

Dolan, E. F. 1991. *Our Poisoned Sky*. New York: Cobblehill Books.

EPA. 2005. *Basic Air Pollution Meteorology*. Accessed January 12, 2008, at www.epa. gov/apti.

Hansen, J. E., et al. 1986. "Climate Sensitivity to Increasing Greenhouse Gases," *Greenhouse Effect and Sea Level Rise: A Challenge for This Generation*, ed., M. C. Barth and J. G. Titus. New York: Van Nostrand Reinhold.

Hansen, J. E., & Lebedeff, F. 1989. "Greenhouse Effect of Chlorofluorocarbons and Other Trace Gases." *Journal of Geophysical Research*, 94 (November), pp. 16, 417–421.

Time. 1998. "Global Warming: It's Here . . . and Almost Certain to Get Worse." August 24.

USA TODAY. 1997. "Global Warming: Politics and Economics Further Complicate the Issue," p. A-1, 2, December 1, 1997.

USEPA. (1995). *The Probability of Sea Level Rising*. Washington, DC: Environmental Protection Agency.

III

WEATHER AND CLIMATE

The Pharisees also with the Sadducees came and tested him, asking that he would show them a sign from heaven. He answered and said unto them, When it is evening ye say, It will be fair weather today for the sky is red and lowering. Oh ye hypocrites, ye can discern the face of the sky, but can ye not discern the signs of the times.

—Matthew 16:1–4

8

The Science of Weather and Climate

One can make a day of any size, and regulate the rising and setting of his sun and the brightness of the shining.

—John Muir, 1875

Mean Weather
Intermittent rain, I've learned,
Which forecasts tell about,
Is rain that stops when I go in
And starts when I come out.

—Elizabeth Dolan (from *The Breeze*,
Vol. 2, No. 8, September 10, 1945)

Red sky at night, sailor's delight.
Red sky in morning, sailor take warning.

—Weather lore

THE EMINENT METEOROLOGIST Edward Norton Lorenz (1917–2008) once said, "A butterfly flapping its wings in Brazil can cause a tornado in Texas." What the meteorologist was implying is true to a point (and in line with what some critics might say): Because of tiny nuances in Earth's weather patterns, making accurate, long-range weather predictions is extremely difficult.

What is the difference between weather and climate? Some people get these two confused, believing they mean the same thing, but they do not. In this chapter you will gain a clear understanding of the meaning of and difference between the two and also gain basic understanding of the role weather plays in air pollution.

Meteorology: The Science of Weather

As already mentioned, *meteorology* is the science concerned with the atmosphere and its phenomena. The atmosphere is the media into which all air pollution is emitted. The meteorologist observes atmospheric processes such as temperature, density, (air) winds, clouds, precipitation, and other characteristics and endeavors to account for their observed structure and evaluation (weather, in part) in terms of external influence and the basic laws of physics. *Air pollution meteorology* is the study of how these atmospheric processes affect the fate of air pollutants.

Since the atmosphere serves as the medium into which air pollutants are released, the transport and dispersion of these releases are influenced significantly by meteorological parameters. Understanding air pollution meteorology and its influence in pollutant dispersion is essential in air quality planning activities. Planners use this knowledge to help locate air pollution monitoring stations and to develop implementation plans to bring ambient air quality into compliance with standards. Meteorology is used in predicting the ambient impact of a new source of air pollution and to determine the effect on air quality from modifications to existing sources (EPA, 2005).

Weather is the state of the atmosphere, mainly with respect to its effect upon life and human activities; as distinguished from *climate* (the long-term manifestations of weather), weather consists of the short-term (minutes or months) variations of the atmosphere. Weather is defined primarily in terms of heat, pressure, wind, and moisture.

Did You Know?

The difference between weather and climate is a measure of time. Weather is what conditions of the atmosphere are over a short period of time, and climate is how the atmosphere "behaves" over relatively long periods of time (NASA, 2011).

At high levels above the earth, where the atmosphere thins to near vacuum, there is no weather; instead, weather is a near-surface phenomenon. This is evidenced clearly on a day-by-day basis where you see the ever-changing, sometimes dramatic, and often violent weather display.

In the study of air science and, in particular, of air quality, the following determining factors are directly related to the dynamics of the atmosphere, resulting in local weather. These factors include strength of winds, the direction they are blowing, temperature, available sunlight (needed to trigger photochemical reactions, which produce smog), and the length of time since the last weather event (strong winds and heavy precipitation) cleared the air.

Weather is basically the way the atmosphere is behaving, mainly with respect to the effects upon life and human activities. There are a lot of components to weather. Weather includes sunshine, rain, cloud cover, winds, hail, snow, sleet, freezing rain, flooding, blizzards, ice storms, thunderstorms, steady rains from a cold front or warm front, excessive heat, heat waves, and more (NASA, 2011). Weather events (such as strong winds and heavy precipitation) that work to clean the air we breathe are beneficial. However, few people would categorize the weather events such as tornadoes, hurricanes, and typhoons as beneficial. Other weather events have both a positive and negative effect. One such event is El Niño / Southern Oscillation, discussed below.

El Niño / Southern Oscillation

El Niño / Southern Oscillation, or ENSO, is a natural phenomenon that occurs every 2 to 9 years on an irregular and unpredictable basis. El Niño is a warming of the surface waters in the tropical eastern Pacific, which causes fish to disperse to cooler waters and, in turn, causes the adult birds to fly off in search of new food sources elsewhere.

Through a complex web of events, El Niño (which means "the child" in Spanish because it usually occurs during the Christmas season off the coasts of Peru and Ecuador) can have a devastating impact on all forms of marine life.

During a normal year, equatorial trade winds pile up warm surface waters in the western Pacific. Thunderheads unleash heat and torrents of rain. This heightens the east-west temperature difference, sustaining the cycle. The jet stream blows from north Asia to California. During an ENSO year, trade winds weaken, allowing warm waters to move east. This decreases the east-west temperature difference. The jet stream is pulled farther south than normal, picks up storms it would usually miss, and carries them to Canada or California. Warm waters eventually reach South America.

One of the first signs of its appearance is a shifting of winds along the equator in the Pacific Ocean. The normal easterly winds reverse direction and drag a large mass of warm water eastward toward the South American coastline. The large mass of warm water basically forms a barrier that prevents the upwelling of nutrient-rich cold water from the ocean bottom to the surface. As a result, the growth of microscopic algae that normally flourish in the nutrient-rich upwelling areas diminishes sharply, and that decrease has further repercussions. For example, El Niño / Southern Oscillation has been linked to patterns of subsequent droughts, floods, typhoons, and other costly weather extremes around the globe. Take a look at ENSO's effect on the west coast of the United States where the weather pattern has been blamed for west coast hurricanes, floods, and early snowstorms. On the positive side, ENSO typically brings good news to those who live on the east coast of the United States: a reduction in the number and severity of hurricanes.

Note that, in addition to reducing the number and severity of hurricanes, ENSO, as reported in October 1997 by the Associated Press in a new study, also deserves credit for invigorating plants and helping to control the pollutant linked to global warming. Researchers have found that El Niño causes a burst of plant growth throughout the world and this removes carbon dioxide from the atmosphere.

Atmospheric carbon dioxide (CO_2) has been increasing steadily for decades. The culprits are increased use of fossil fuels and the clearing of tropical rainforests. However, during an ENSO phenomenon, global weather is warmer, there is an increase in new plant growth and CO_2 levels decrease.

Not only does ENSO have a major regional impact in the Pacific, its influence extends to other parts of the world through the interaction of pressure, airflow, and temperature effects.

It is a phenomenon that, although not quite yet completely understood by scientists, causes both positive and negative results, depending upon where you live.

Air Masses

An air mass is a vast body of air (a macroscale phenomenon that can have global implications) in which the conditions of temperature and moisture are much the same at all points in a horizontal direction. An air mass takes on the temperature and moisture characteristics of the surface over which it forms and travels, though its original characteristics tend to persist. The processes of radiation, convection, condensation, and evaporation condition the air in an air mass as it travels. Also, pollutants released into an air mass travel and disperse within the air mass. Air masses develop more commonly in some regions than in others. Table 8.1 summarizes air masses and their properties.

TABLE 8.1
Classification of Air Masses

Name	Origin	Properties	Symbol
Arctic	Polar regions	Low temperatures, low specific but high summer relative humidity, the coldest of the winter air masses	A
Polar continental*	Subpolar continental areas	Low temperatures (increasing with southward movement), low humidity, remaining constant	cP
Polar maritime	Subpolar area and arctic region	Low temperatures increasing with movement, higher humidity	mP
Tropical continental	Subtropical high-pressure land areas	High temperatures, low moisture content	cT
Tropical maritime	Southern borders of oceanic subtropical, high-pressure areas	Moderate high temperatures, high relative and specific humidity	mT

* The name of an air mass, such as polar continental, can be reversed to continental polar, but the symbol, cP, is the same for either name.

Source: EPA, 2005.

When two different air masses collide, a *front* is formed. A front is not a sharp wall but a zone of transition that is often several miles wide. Four frontal patterns—warm, cold, occluded, and stationary—can be formed by air of different temperatures. A *warm front* (see Figure 8.1) marks the advance of a warm air mass as it rises up over a cold one. A *cold front* (see Figure 8.2) marks the line of advance of a cold air mass from below, as it displaces a warm air mass.

When cold and warm fronts merge (the cold front overtaking the warm front), *occluded fronts* form. Occluded fronts can be called cold front or warm front occlusions. But, in either case, a colder air mass takes over an air mass that is not as cold.

The last type of front is the stationary front. As the name implies, the air masses around this front are not in motion. A stationary front can cause bad weather conditions that persist for several days. Figure 8.3 illustrates vertical profiles of various fronts.

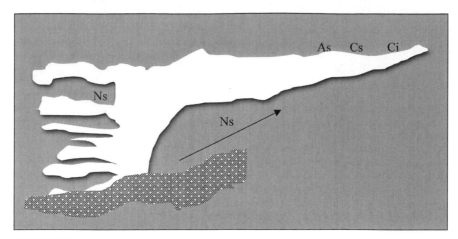

FIGURE 8.1
An idealized warm front showing the succession of cloud types: Ns—nimbostratus,
As—altostratus, Cs—Cirrostratus, and Ci—cirrus.

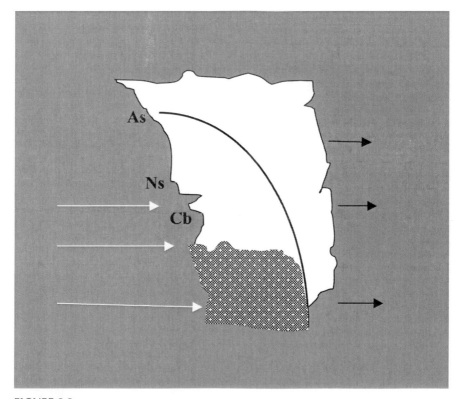

FIGURE 8.2
An idealized cold front with As—altostratus, Ns—nimbostratus, and Cb—cumulonimbus
cloud species.

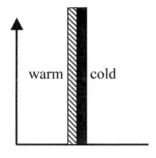 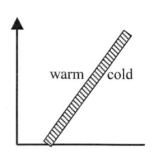

1. Stationary Front   2. Warm Front

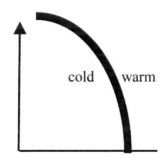 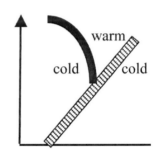

3. Cold Front   4. Occluded Front

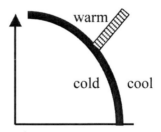

5. Occluded Fronts

   warm

                             cold

FIGURE 8.3
Vertical profiles of fronts.

Optical Phenomena

Have you ever looked up into the sky and seen eleven suns? Have you been at sea and witnessed the towering, spectacular Fata Morgana? (Do you even know what Fata Morgana is? If not, hold on, we'll get to it shortly.) How about a "glory"—have you ever seen one? Or how about the albino rainbow—have you seen one lately? Do you know what these are? They will be described shortly.

Normally, when we look up into the sky, we see what we expect to see: an ever-changing backdrop of color, with dynamic vistas of blue sky; white, puffy clouds; gray storms; and gold and red sunsets. On some occasions, however, when atmospheric conditions are just right, we can look up at the sky or out upon the horizon and see the strange phenomena (or lights in the sky) mentioned above. What causes these momentary wonders?

Because earth's atmosphere is composed of gases (air)—it is actually a sea of molecules. These molecules of air scatter the blue, indigo, and violet shorter wavelengths of light more than the longer orange and red wavelengths, which is why the sky appears blue.

What are wavelengths of light? Simply put, a wavelength of light actually refers to the electromagnetic spectrum. The portion of the spectrum visible to the human eye falls between the infrared and ultraviolet wavelengths. As mentioned, the colors that make up the visible portion of the electromagnetic spectrum are commonly abbreviated by the acronym ROY G BIV (Red, Orange, Yellow, Green, Blue, Indigo, and Violet).

The word *light* is commonly given to visible electromagnetic radiation. However, only the frequency (or wavelength) distinguishes visible electromagnetic radiation from the other portions of the spectrum.

Let's get back to the sky, that is, to looking up into the sky. Have you ever noticed that right after a rain shower, how dark a shade of blue the sky appears? Have you looked out upon the horizon at night or in the morning and noticed that the sun's light gives off a red sky? This phenomenon is caused by sunlight passing through large dust particles, which scatter the longer wavelengths. Have you ever noticed that fog and cloud droplets, with diameters larger than the wavelength of light, scatter all colors equally and make the sky look white? Maybe you have noticed that fleeting greenish light that appears just as the sun sets? It occurs because different wavelengths of light are *refracted* (bent) in the atmosphere by differing amounts. Because green light is refracted more than red light by the atmosphere, green is the last to disappear.

What causes rainbows? A rainbow is really nothing more than an airborne prism. When sunlight enters a raindrop, refraction and reflection take place, splitting white light into the spectrum of colors from red to blue and making a rainbow. Because the light is reflected inside the raindrops, rainbows appear on the opposite side of the sky to the sun.

Earlier, a "glory" was mentioned. Interactions of light waves can produce a glory, an optical phenomenon. A glory often appears as an iconic saint's halo (a series of colored rings) about the head of the observer, which is produced by a combination of diffraction, reflection, and refraction (backscattered light) toward its source by a cloud of uniformly sized water droplets. For example, if you were standing on a mountain, with the sun to your back, you may cast a shadow on the fog in the valley. Your shadow may appear to be surrounded by colored halos—a glory. The glory is caused by light entering the edges of tiny droplets and being returned in the same direction from which it arrived. These light waves interfere with each other, sometimes canceling out and sometimes adding to each other.

A real-life example of a group in 1893 at Gausta Mountain, Norway, that saw well-defined glories was reported in *Nature* (48:391, 1893): "We mounted to the flagstaff in order to obtain a better view of the scenery, and there we at once observed in the fog, in an easterly direction, a double rainbow forming a complete circle and seeming to be 20 to 30 feet distance from us. In the middle of this we all appeared as black, erect, and nearly life-size silhouettes. The outlines of the silhouettes were so sharp that we could easily recognize the figures of each other, and every movement was reproduced. The head of each individual appeared to occupy the centre of the circle, and each of us seemed be standing on the inner periphery of the rainbow."

Did You Know?

Similar to a glory, a heiligenschein ("holy light") is a bright halo of light that appears around the shadow of the observer's head. As with a glory, the observer sees only the light surrounding his or her own head, not around those of any companions.

Why do we sometimes see multiple suns? Reflection and refraction of light by ice crystals can create bright halos in the form of arcs, rings, spots, and pillars. Mock suns (sun dogs) may appear as bright spots 22° or 46° to the left or right of the sun. A one-sided mock sun was observed by several passengers aboard the ship *Fairstar* in the Sea of Timor on June 15, 1965. "About 10 minutes before sunset, Mrs. N. S. noticed to her surprise a second sun, much less bright than the real one, somewhat to the left, but at the same height above the horizon. There was nothing at the symmetrical point to the right" ("Fata Morgana," 1966). Sun pillars occur when ice crystals act as mirrors, creating a bright column of light extending above the sun. Such a pillar of bright light may be visible even when the sun has set.

What is the Fata Morgana? It is an illusion, a mirage, that often fools sailors into seeing mountain ranges floating over the surface of the ocean. Henry Wadsworth Longfellow had his own take on Fata Morgana, which basically explains the essence of the phenomenon.

Fata Morgana

O sweet illusions of song
That tempt me everywhere,
In the lonely fields, and the throng
Of the crowded thoroughfare!
I approach and ye vanish away,
I grasp you, and ye are gone;
But over by night and by day,
The melody soundeth on.
As the weary traveler sees
In desert or prairie vast,
Blue lakes, overhung with trees
That a pleasant shadow cast;
Fair towns with turrets high,
And shining roofs of gold,
That vanish as he draws nigh,
Like mists together rolled—
So I wander and wander along,
And forever before me gleams
The shining city of song,
In the beautiful land of dreams.
But when I would enter the gate
Of that golden atmosphere,
It is gone, and I wonder and wait
For the vision to reappear.

Then there is Socrates: Didst thou never espy a Cloud in the sky,
Which a centaur or leopard might be?
Or a wolf or a cow?
Strepsiades: Very often, I vow:
And show me the cause, I entreat.
Socrates: Why, I tell you that these become just what they please. . . .

—Aristophanes, *The Clouds*, ca. 420 B.C.E.
as translated by Benjamin Bickley Rogers

The flagship account of the Fata Morgana comes to us from friar Antonio Minasi who described a Fata Morgana seen across the Straits of Messina in 1773. "In Italian legend, Morgan Le Fay, or, in Italian, Fata Morgan, falls in love with a mortal youth and gives him the gift of eternal life in return for her love; when he becomes restless and bored with captivity, she summons up fairy spectacles for his entertainment" (Minasi, 1773).

An albino rainbow (white rainbow or fogbow) is an eerie phenomenon that can be seen only on rare occasions in foggy conditions. They form when the sun or moon shine on minute droplets of water suspended in the air. The fog droplets are so small that the usual prismatic colors of the rainbow merge together to form a white arc opposite the sun or moon (Corliss, 1983).

Atmospheric phenomena (lights in the sky) are real, apparent, and sometimes visible. Awe inspiring as they are, their significance—their actual existence—is based on physical conditions that occur in our atmosphere.

References and Recommended Reading

American Meteor Society. 2001. *Definition of Terms by the IAU Commission 22, 1961.* Accessed October 10, 2010, at www.amsmeteors.org/define.html.

Corliss, W. R. 1983. *Handbook of Unusual Natural Phenomena.* New York: Anchor Press.

EPA. 2005. *Basic Air Pollution Meteorology.* APTI Course SI: 409. Accessed January 8, 2008, at www.epa.gov/apt.

EPA. 2007. Basic Concepts in Environmental Sciences: Modules 1 and 2. Accessed December 30, 2007, at www.epa.gov/apti/bces/home/index.htm.

"Fata Morgana." (1966). In *Weather,* 21: 250.

Heinlein, R. A. 1973. *Time Enough for Love,* New York: G.P. Putnum's Sons.

Hesketh, H. E. 1991. *Air Pollution Control: Traditional and Hazardous Pollutants.* Lancaster, PA: Technomic Publishing Company.

LiveScience. 2010. *Large Swaths of Earth Drying Up, Study Suggests.* Accessed October 13, 2010, at www.livescience.com/8755-large-swaths-earth-drying-study-suggests.html.

Minasi, A. 1773. "Dissertazioni, Rome." In Warner, M. 1999. The Tanner Lectures on Human Values at Yale University.

NASA. 2007. *Pascal's Principle and Hydraulics.* Accessed December 29, 2007, at www.grc.nasa.gov/WWW/k-12/WindTunnel/Activities/Pascals_principle.html.

NASA 2011. *What's the Difference between Weather and Climate?* Accessed July 9, 2011, at www.nasa.gov/mission_pages/noaa-n/climate/climate_weather.html.

National Academy of Sciences. 1962. *Water Balance in the U.S.* National Research Council Publication 100-B.

Shipman, J. T., Adams, J. L., & Wilson, J. D. 1987. *An Introduction to Physical Science.* Lexington, MA: D. C. Heath and Company.

Spellman, F. R., 2007. *The Science of Water,* 2nd ed. Boca Raton, FL: CRC Press.

Spellman, F. R., 2009. *The Science of Air*, 2nd ed. Boca Raton FL: CRC Press.

Spellman, F. R., & Whiting, N. 2006. *Environmental Science & Technology: Concepts and Applications.* Rockville, MD: Government Institute.

USGS. 2010. *The Water Cycle: Evapotranspiration.* Accessed October 7, 2010, at http://ga.water.usgs.gov/edu/ watercycleevapotranspiration.html.

9

Microclimates

Nothing that is can pause or stay;
The moon will wax, the moon will wane,
The mist and cloud will turn to rain,
The rain to mist and cloud again,
Tomorrow be today.

—Henry Wadsworth Longfellow

When morning fog clears away quickly, expect a sunny day.

—Weather lore

WHEN WE THINK ABOUT CLIMATE, we are generally referring to overall or generalized weather conditions at a particular place or region over a period of time. In addition to precipitation and temperature, climates have been classified into zones by vegetation, moisture index, and even measures of human discomfort. Using the general climate zone names allows us to differentiate between a particular climate (with its specific climatic conditions) in respect to another climate with differing conditions. When geographical patterns in the weather occur again and again over a long period, they can all be used to define the climate of a region. Some climate zones are known as the hot climates, which include desert, tropical continental, tropical monsoon, tropical marine, or equatorial types. Warm climates include west coast (Mediterranean) and warm east coast. Another category includes the cool climates such as cold desert, west coast (cool), cool temperate interior, and cool temperate east coast types. Finally, there are the mountain and the cold climate categories of cold continental and polar or tundra. Each of these different climate types is differentiated from each other. However, they all have one major feature in common: They are large-scale regional climates (with variations), occurring at various places throughout the world. They consider only the broad similarities between a particular climate at various locations worldwide; local differences are ignored and boundaries are approximate.

What factors determine the variations of climate over the surface of the earth? The primary factors are (1) the effect of latitude and the tilt of the earth's axis on the plane of the orbit about the sun, (2) the large-scale movements of different wind belts over the earth's surface, (3) the temperature difference between land and sea, (4) the contours of earth's surface, and (5) the location of the area in relation to ocean currents.

What factors determine how climates are distributed? When considering climate distribution, remember that the world does not fall into compartments. The globe is a mosaic of many different types of climate, just as it is a mosaic of numerous types of ecosystems. The complexity of the distribution of land and sea and the consequent complexity of the general circulation of the atmosphere have a direct effect on the distribution of the climate.

Types of Microclimates

What is a microclimate? To answer this question, we must first talk about scale. For example, let's take a look at flow of air within a very small environment: the emission of smoke from a chimney. This flow represents one of the smallest spatial subdivisions of atmospheric motion, or microscale weather. On a more realistic, but still relatively small scale, we must consider the geographical, biological, and man-made features that make local climate different from the general climate. This local climatic pattern is called a *microclimate*.

What are the elements or conditions that cause local or microclimates? Location, location, location—and local conditions—are the main ingredients making up a microclimate. Let's look at one example.

Large inland lakes moderate temperature extremes and climatic differences between the windward and lee sides. For example, Seattle, on the windward side of Lake Washington, and Bellevue, on the lee side only about 9 miles east, have microclimatic differences (although modest) between the two cities. These microclimatic differences exist in temperature fluctuations, precipitation levels, wind speed, and relative humidity.

Even more dramatic differences can be seen in such parameters when a comparison is made between a city such as Milwaukee, on the windward side of Lake Michigan, and Grand Haven, on the lee side, only 85 miles east. Other examples of microclimates can be found in the following areas:

- Near the ground
- Over open land areas
- In woodlands or forested areas
- In valley regions
- In hillside regions
- In urban areas
- In seaside locations

In the following sections we take a closer look at these microclimates: at their nature, causative factors, and geographical/topographical locations.

Microclimates Near the Ground

Nowhere in the atmosphere are climatic differences as distinct as they are near the ground. For instance, when you go to the beach on a warm summer day, you no doubt have noticed that the grass and water are much cooler to your feet than the sand. So, you may ask, what is it about this area near the ground that produces a microclimate with such major differences?

It's the interface (or activity zone) between the atmosphere and the ground surface (sandy shore) that causes the stark difference in temperature variability. Energy is reaching the sandy beach from the sun and from the atmosphere (though to a much lesser extent). The energy is either reflected and then returned to the atmosphere in a different form or is absorbed and stored in the sandy surface as heat.

Ground level energy absorption is very sensitive to the nature of the ground surface. Ground surface color, wetness, cover (vegetation), and topography are conditions that all affect the interaction between the ground and the atmosphere. Consider a snow-covered ground, for example. Clean snow reflects solar radiation, so the surface remains cool and the snow fails to melt. However, dirty snow absorbs more radiation, heats up, and is likely to melt. If the snowy area is shielded by vegetation, the vegetation, too, may protect the snow from the heat of the sun.

Microclimate Over Open Land Areas

Many different properties of ground layer or soil type influence conditions in the thin layer of atmosphere just above the ground. Light-colored soils do not absorb energy as efficiently as do organically rich darker soils. Another important factor is soil moisture. Wet soils are normally dark, but moist soil (because water has a large heat capacity) requires a great deal of energy to raise its temperature. A moist soil warms up more slowly than a dry one.

Soil is a heterogeneous mixture of various particles. In between the soil particles is a large amount of air—air that is a poor conductor of heat. The larger the amount of air between the soil particles, the slower the heat transfers through the soil. As demonstrated in our example of the sandy beach, on a hot sunny day the heat is trapped in the upper layers, so the surface layers warm up more rapidly and become extremely hot. Water conducts heat more readily than air, so soils that contain some moisture are able to transmit warmth away from the surface more easily than dry soils. This is not always the case, however. If the soil contains too much water, the large heat capacity of the water will prevent the soil from warming despite heat being conducted from the surface.

Microclimates in Woodlands or Forested Areas

When making microclimate comparisons between open land areas and forested areas (commonly referred to as a forest climate), the differences are apparent. Forested areas, for example, are generally warmer in winter than the open areas, while open land is warmer in summer than forested areas. The forest climate has reduced wind speeds, while the open land area has higher wind speeds. The forest climate has higher relative humidity, while the open area has lower relative humidity. In the forest climate, water storage capacity is higher and evaporation rates are lower, while in the open land area water storage capacity is lower and evaporation is higher.

Microclimates in Valley and Hillside Regions

Heavy, cold air flows downhill, forming cold pockets in valleys. Frost is much more common there, so orchards of apples and oranges and vines of grapes are planted on hillsides to ensure frost drainage when cold spells come.

Probably the best way in which to describe the microclimate in a typical valley region is to compare and contrast it with a hillside environment. In a typical valley region, the daily minimum temperature is much lower than that in a hillside area. The daily and annual temperature range for a valley is much larger than that of a hillside area. In a valley region, more frost occurs than in a hillside region. Windspeed at night is lower in a valley than on a hillside, and morning fog is more prevalent and lasts longer in a valley region.

Did You Know?

Urban areas have added roughness features and different thermal characteristics due to the presence of man-made elements. The thermal influence dominates the influence of the frictional components. Building materials such as brick and concrete absorb and hold heat more efficiently than soil and vegetation found in rural areas. After the sun sets, the urban area continues to radiate heat from buildings, paved surfaces, and so on. Air warmed by the urban complex rises to create a dome over the city. It is called the *heat island effect*. The city emits heat all night. Just when the urban area begins to cool, the sun rises and begins to heat the urban complex again. Generally, city areas never revert to stable conditions because of the continual heating that occurs (EPA, 2005).

Microclimates in Urban Areas

The microclimate in an urban area as compared to that of the countryside is usually quite obvious. A city, for example, is usually characterized by having haze and smog, higher temperatures, lower wind speed, and reduced radiation. The countryside, on the other hand, is characterized by clear air, lower temperatures, and high wind speeds and radiation.

These different microclimatic conditions should come as no surprise to anyone, especially when you consider what happens when a city is built. Instead of a mixture of soil or vegetation, the surface layer is covered with concrete, brick, glass, and stone surfaces ranging to heights of several hundred feet. These materials have vastly different physical properties from soil and trees. They shed and carry away water, absorb heat, block and channel the passage of winds, and present albedo levels significantly different from those of the natural world. All of these factors (and more) work to alter the climate conditions in the area.

Microclimates in Seaside Locations

The major climatic feature associated with seaside locations is the sea breeze. Sea breezes are formed by the different responses to heating of water and land. For example, if we have a bright, sunny morning with little wind, the ground surface warms rapidly as it absorbs short-wave radiation. Most of this heat is retained at the surface, although some will be transferred through the soil. As a result, the temperature of the ground surface increases and some of the heat warms the air above. When the sun sets, the surface starts to cool rapidly because there is little store of heat in the soil. Thus, we find that land surfaces are characterized by high day (and summer) temperatures and low night (and winter) temperatures.

Now let's take a look at the response of the sea, which is very different. Solar energy (sunshine) is able to penetrate through the water to a certain level. Much solar energy has to be absorbed to raise its temperature. Through wave action and convection, the warm surface water is mixed with cooler deeper water. With enough solar energy and time, the top several feet of water form as an active layer where temperature change is slow. Slight warming occurs during the day and slight cooling at night. This means that the sea is normally cooler than the land by day and warmer by night.

The higher temperature over the land by day generates a weak low-pressure area. As this intensifies during daytime heating, a flow of cool, more humid air spreads inland from the sea, gradually changing in strength and direction during the day. At night the reverse occurs, with circulation of air from the cooler land to the warmer sea, though as the temperature difference is usually less, the land breeze is weak. Even large lakes can show a breeze system of this nature.

References and Recommended Reading

EPA. 2005. *Basic Air Pollution Meteorology.* APTI Course SI: 409. Accessed January 9, 2008, at www.epa.gov/apti.

Spellman, F. R., & Whiting, N. 2006. *Environmental Science and Technology: Concepts and Applications,* 2nd ed. Rockville, MD: Government Institutes.

IV

ATMOSPHERIC DYNAMICS

A sun shiny shower won't last half an hour.

—Weather lore

10

The Atmosphere in Motion

The winds wander, the snow and rain and dew fall, the earth whirls—all but to prosper a poor lush violet.

—John Muir, 1913

O Wild West Wind, thou breath of Autumn's being
Thou from whose unseen presence the leaves dead
Are driven like ghosts from an enchanter fleeing,
Yellow, and black, and pale, and hectic red,
Pestilence-stricken multitudes! O thou
Who chariotest to their dark wintry bed
The wingèd seeds, where they lie cold and low,
Each like a corpse within its grave, until
Thine azure sister of the Spring shall blow
Her clarion o'er the dreaming earth, and fill
(Driving sweet buds like flocks to feed in air)
With living hues and odours plain and hill;
Wild Spirit, which art moving everywhere;
Destroyer and preserver; hear, O hear!

—Percy Bysshe Shelley, *Ode to the West Wind*, 1819

HAVE YOU EVER WONDERED WHY the earth's atmosphere is in perpetual motion? Probably not—but it is. It must be in a state of perpetual motion because it constantly strives to eliminate the constant differences in temperature and pressure between different parts of the globe. How are these differences eliminated or compensated for? By its motion, the earth's atmosphere produces winds and storms. In this chapter the horizontal movements that transfer air around the globe are considered.

Global Air Movement

Basically, winds are the movement of the earth's atmosphere, which by its weight exerts a pressure on the earth that we can measure using a barometer. Winds are often confused with air currents, but they are different. Wind is the horizontal movement of air or motion along the earth's surface. Air currents, on the other hand, are vertical air motions collectively referred to as updrafts and downdrafts.

Throughout history, man has been both fascinated by and frustrated by winds. Man has written about winds almost from the time of the first written word. For example, Herodotus (and later Homer and many others) wrote about winds in his *The Histories*. Wind has had such an impact upon human existence that we have given winds names that describe a particular wind, specific to a particular geographical area. Table 10.1 lists some of these winds, their colorful names, and the region where they occur. Some of these names are more than just colorful—the winds are actually colored. For example, the *harmattan* blows across the Sahara filled with red dust; mariners called this red wind the "sea of darkness."

TABLE 10.1
Assorted Winds of the World

Wind Name	Location
Aajej	Morocco
Alm	Yugoslavia
Biz roz	Afghanistan
Haboob	Sudan
Imbat	North Africa
Datoo	Gibraltar
Nafhat	Arabia
Besharbar	Caucasus
Samiel	Turkey
Ṭsumuji	Japan
Brickfielder	Australia
Chinook	America
Williwaw	Alaska

Earth's Atmosphere in Motion

To state that earth's atmosphere is constantly in motion is to state the obvious. Anyone observing the constant weather changes around them is well
aware of this phenomenon. Although obvious, the importance of the dynamic
state of our atmosphere is much less obvious.

As mentioned, the constant motion of earth's atmosphere (air movement)
consists of both horizontal (wind) and vertical (air currents) dimensions. The
atmosphere's motion is the result of thermal energy produced from the heating of the earth's surface and the air molecules above. Because of differential
heating of the earth's surface, energy flows from the equator poleward.

Even though air movement plays the critical role in transporting the energy of the lower atmosphere, bringing the warming influences of spring
and summer and the cold chill of winter, the effects of air movements on
our environment are often overlooked, even though wind and air currents
are fundamental to how nature functions. All life on earth has evolved with
mechanisms dependent on air movement: Pollen is carried by winds for plant
reproduction; animals sniff the wind for essential information; wind power
was the motive force that began the earliest stages of the industrial revolution.
Now we see the effects of winds in other ways, too: Wind causes weathering
(erosion) of the earth's surface; wind influences ocean currents; air pollutants and contaminants such as radioactive particles transported by the wind
impact our environment.

Causes of Air Motion

In all dynamic situations, forces are necessary to produce motion and changes
in motion—winds and air currents. The air of the atmosphere (made up of
various gases) is subject to two primary forces: (1) gravity and (2) pressure
differences from temperature variations.

Gravity (gravitational forces) holds the atmosphere close to the earth's
surface. Newton's law of universal gravitation states that everybody in the
universe attracts another body with a force equal to:

$$F = G \, \frac{m_1 m_2}{r^2} \tag{10.1}$$

where F = force, m_1 and m_2 = the masses of the two bodies, G = universal
constant of $6.67 \times 10^{-11} \, N \times m^2/kg^2$, and R = distance between the two bod-

ies. *Important point:* The force of gravity decreases as an inverse square of the distance between the two bodies.

Thermal conditions affect density, which in turn causes gravity to affect vertical air motion and planetary air circulation. This affects how air pollution is naturally removed from the atmosphere.

Although forces in other directions often overrule gravitational force, the ever-present force of gravity is vertically downward and acts on each gas molecule, accounting for the greater density of air near the earth.

Atmospheric air is a mixture of gases, so the gas laws and other physical principles govern its behavior. The pressure of a gas is directly proportional to its temperature. Pressure is force per unit area ($P = F/A$), so a temperature variation in air generally gives rise to a difference in pressure of force. This difference in pressure resulting from temperature differences in the atmosphere creates air movement—on both large and local scales. This pressure difference corresponds to an unbalanced force, and when a pressure difference occurs, the air moves from a high- to a low-pressure region.

In other words, horizontal air movements (called advective winds) result from temperature gradients, which give rise to density gradients and, subsequently, pressure gradients. The force associated with these pressure variations (pressure gradient force) is directed at right angles to (perpendicular to) lines of equal pressure (called isobars) and is directed from high to low pressure.

Look at Figure 10.1. The pressures over a region are mapped by taking barometric readings at different locations. Lines drawn through the points (locations) of equal pressure are called isobars. All points on an isobar are of equal pressure, which means no air movement along the isobar. The wind direction is at right angles to the isobar in the direction of the lower pressure. In Figure 10.1, notice that air moves down a pressure gradient toward a lower isobar like a ball rolls down a hill. If the isobars are close together, the pressure gradient force is large, and such areas are characterized by high wind speeds. If isobars are widely spaced (see Figure 10.1), the winds are light because the pressure gradient is small.

Did You Know?

Air pressure at any location, whether it is on the earth's surface or up in the atmosphere, depends on the weight of the air above. Imagine a column of air. At sea level, a column of air extending hundreds of kilometers above sea level exerts a pressure of 1,013 millibars (mb). But, if you travel up the column to an altitude of 4.4 km (18,000 feet), the air pressure would be roughly half, or approximately 506 mb.

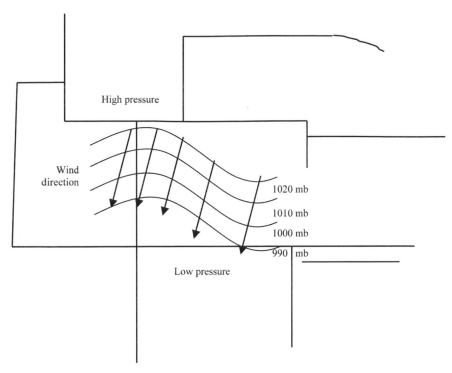

FIGURE 10.1
Isobars drawn through locations having equal atmospheric pressures. The air motion, or wind direction, is at right angles to the isobars and moves from a region of high pressure to a region of low pressure.

Localized air circulation gives rise to thermal circulation (a result of the relationship based on a law of physics whereby the pressure and volume of a gas is directly related to its temperature). A change in temperature causes a change in the pressure and/or volume of a gas. With a change in volume comes a change in density, since $P = m/V$, so regions of the atmosphere with different temperatures may have different air pressures and densities. As a result, localized heating sets up air motion and gives rise to thermal circulation. To gain understanding of this phenomenon, consider Figure 10.2.

Once the air has been set into motion, secondary forces (velocity-dependent forces) act. These secondary forces are (1) earth's rotation (Coriolis force) and (2) contact with the rotating earth (friction). The *Coriolis force*, named after its discoverer, French mathematician Gaspard Coriolis (1772–1843), is the effect of rotation on the atmosphere and on all objects on the earth's surface; simply, it is air deflection. In the northern hemisphere,

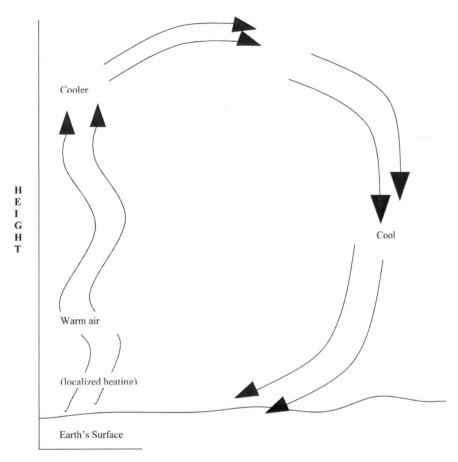

FIGURE 10.2
Thermal circulation of air. Localized heating, which causes air in the region to rise, initiates the circulation. As the warm air rises and cools, cool air near the surface moves horizontally into the region vacated by the rising air. The upper, still cooler, air then descends to occupy the region vacated by the cool air.

it causes moving objects and currents to be deflected to the right (rightward effect); in the southern hemisphere, it causes deflection to the left because of the earth's rotation. The Coriolis effect is stronger northward because the latitudinal circles there describe much wider variations in the measurements of the circumferences of the circles that are right next to each other than those circumferences of the latitudinal circles at or around the equator. Air, in large-scale north or south movements, appears to be deflected from its expected path. That is, air moving poleward in the northern hemisphere appears to be deflected toward the east; air moving southward appears to be deflected toward the west.

Figure 10.3 illustrates the Coriolis effect on a propelled particle (analogous to the apparent effect of an air mass flowing from point A to point B). From Figure 10.3, the action of the earth's rotation on the air particle as it travels north over the earth's surface, as earth rotates beneath it from east to west,

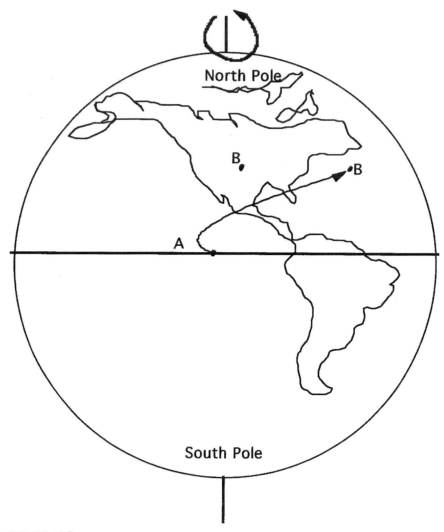

FIGURE 10.3
The effect of the earth's rotation on the trajectory of a propelled particle.

can be seen. Projected from point A to point B, the particle will actually reach Point B because as it is moving in a straight line (deflected), the earth rotates east to west beneath it.

Did You Know?

The Coriolis effect is present not only in the atmosphere causing its characteristic deviations and in toilet contents swirling down the drain, but also in the meandering of streams, the right-hand erosion of stream banks in the northern hemisphere, the wearing of the right front brakes on a car, the wearing of the right-hand rail on train tracks where a train travels in one direction only, and the rotation of air within a tornado and a hurricane.

Friction (drag) can also cause the deflection of air movements. This friction (resistance) is both internal and external. The friction of its molecules generates internal friction. Friction is also generated when air molecules run into each other. External friction is caused by contact with terrestrial surfaces. The magnitude of the frictional force along a surface is dependent on the air's magnitude and speed, and the opposing frictional force is in the opposite direction of the air motion.

Did You Know?

Friction, one of the major forces affecting the wind, comes into play near the earth's surface and continues to be a factor up to altitudes of about 500 to 1,000 meters. This section of the atmosphere is referred to as the planetary or atmospheric boundary layer. Above this layer, friction no longer influences the wind.

Local and World Air Circulation

Air moves in all directions, and these movements are essential for us here on earth: Vertical air motion is essential in cloud formation and precipitation. Horizontal air movement near the earth's surface produces winds.

Wind is an important factor in human comfort, especially affecting how cold we feel. A brisk wind at moderately low temperatures can quickly make us uncomfortably cold. Wind promotes the loss of body heat, which aggravates the chilling effect, expressed through wind chill factors in the winter (see Table 10.2) and the heat index in the summer (see Table 10.3). These

TABLE 10.2
Wind Chill Chart

Wind MPH	Temperature (Degrees Fahrenheit)										
	30	25	20	10	5	0	−5	−10	−15	−20	−25
5	25	19	13	1	−5	−11	−16	−22	−28	−34	−40
10	21	15	9	−4	−10	−16	−22	−28	−35	−41	−47
15	19	13	6	−7	−13	−19	−26	−32	−39	−45	−51
20	17	11	4	−9	−15	−22	−29	−35	−42	−48	−55
25	16	9	3	−11	−17	−24	−31	−37	−44	−51	−58
30	15	8	1	−12	−19	−26	−33	−39	−46	−53	−60
35	14	7	0	−14	−21	−27	−34	−41	−48	−55	−62
40	13	6	−1	−15	−22	−29	−36	−43	−50	−57	−64
45	12	5	−2	−16	−23	−30	−37	−44	−51	−58	−65
50	12	4	−3	−17	−24	−31	−38	−45	−52	−60	−67
55	11	4	−3	−18	−25	−32	−39	−46	−54	−61	−68
60	10	3	−4	−19	−26	−33	−40	−48	−55	−62	−69

Note: Gray cells indicate frostbite occurs in 15 minutes or less.

Source: USA Today (www.usatoday.com/weather/winter/windchill/windchill/wind-chill-chart.htm).

two scales describe the cooling effects of wind on exposed flesh at various temperatures.

Local winds are the result of atmospheric pressure differences involved with thermal circulations because of geographic features. Land areas heat up more quickly that do water areas, giving rise to a convection cycle. As a result, during the day, when land is warmer than the water, we experience a lake or sea breeze.

At night, the cycle reverses. Land loses its heat more quickly than water, so the air over the water is warmer. The convection cycle sets to work in the opposite direction and a land breeze blows.

In the upper troposphere (above 11 to 14 kilometers, west to east flows) are very narrow fast-moving bands of air called jet streams. Jet streams have significant effects on surface airflows. When jet streams accelerate, divergence of air occurs at that altitude. This promotes convergence near the surface and the formation of cyclonic motion. Deceleration causes convergency aloft and subsidence near the surface, causing an intensification of high-pressure systems.

Jet streams are thought to result from the general circulation structure in the regions where great high- and low-pressure areas meet.

TABLE 10.3
Heat Index Chart (Temperature and Relative Humidity)

RH (%)	Temperature (°F)													
	90	91	92	93	94	95	96	97	98	99	100	101	102	103
90	119	123	128	132	137	141	146	152	157	163	168	174	180	186
85	115	119	123	127	132	136	141	145	150	155	161	166	172	178
80	112	115	119	123	127	131	135	140	144	149	154	159	164	169
75	109	112	115	119	122	126	130	134	138	143	147	152	156	161
70	106	109	112	115	118	122	125	129	133	137	141	145	149	154
65	103	106	108	111	114	117	121	124	127	131	135	139	143	147
60	100	103	105	108	111	114	116	120	123	126	129	133	136	140
55	98	100	103	105	107	110	113	115	118	121	124	127	131	134
50	96	98	100	102	104	107	109	112	114	117	119	122	125	128
45	94	96	98	100	102	104	106	108	110	113	115	118	120	123
40	92	94	96	97	99	101	103	105	107	109	111	113	116	118
35	91	92	94	95	97	98	100	102	104	106	107	109	112	114
30	89	90	92	93	95	96	98	99	101	102	104	106	108	110

Note: Exposure to full sunshine can increase HI values by up to 15°F.

Source: Weather Images (www.weatherimages.org/data/heatindex.html).

Blowing in the Wind . . .

Wind energy conversion—the leading mechanically based renewable energy for much of human history—has been around for thousands of years. It's a technology that has been reinvented numerous times, a proven technology with many advantages for the consumer. Modern wind farms demonstrate that wind turbines are a viable alternative to fossil fuel energy production.

Since cost and capacity factors are so different between wind-generated energy and fossil fuel–generated energy, low installed-cost-per-kilowatt figures for wind turbines are somewhat misleading because of the low capacity factor of wind turbines relative to coal and other fossil-fueled power plants. (Note: "Capacity factor" is the ratio of actual energy produced by a power plant to the potential energy produced if the plant operated at rated capacity for a full year.) Capacity factors of successful wind farm operations range from 0.20 to 0.35. Fossil-fuel power plants have factors of more than 0.50, and some of the new gas turbines reach over 0.60.

"Capacity factor" and the difference between low- and high-capacity production is also misleading. Wind conversion capacity is flexible; production levels vary with the density of the wind resource. More importantly, the wind resource is constant for the life of the machine—not subject to cost manipulation or cost increases. Fossil fuels as energy sources are popular with investors because many of the risks are passed on to consumers. When fossil fuel shortages occur, the investors can raise their prices—causing an *increase* in revenues for investors. In a nasty twist for the consumer, investors in fossil fuel energy production are *rewarded* for (1) speeding the depletion of a nonrenewable resource or (2) not investing enough of their profits in support infrastructure, which drives up prices (think California in 2000–2001). This seeming advantage for wind conversion technology thus becomes a barrier to investment: If big oil, coal, or gas companies could charge consumers for the wind, wind power development would have been a done deal long ago.

The cost of energy from larger electrical output wind turbines used in utility-interconnected or wind farm applications dropped from more than $1.00 per kilowatt-hour (kWh) in 1978 to under $0.05 per kWh in 1998, and it dropped to $0.025 per kWh when new large wind plants came on line in 2001 and 2002. Hardware costs have dropped below $800 per installed kilowatt, lower than the capital costs of almost every other type of power plant.

Wind energy soon will be the most cost effective source of electrical power and perhaps has already achieved this status. The actual life cycle cost of fossil fuels (starting with coal mining and fuel extraction, including transport and use technology, and factoring in environmental impact and political costs) is not really known, but it is certainly far higher than the current wholesale rates—and has been loaded squarely on the shoulders of consumers by the energy industry. From strictly a fuel-cost perspective, since fossil fuel re-sources are non-renewable, the eventual depletion of these energy sources will entail rapid escalations in price. Add to this the environmental and politi-cal costs of fossil fuel use, and the increased awareness of the public to these issues, and fossil fuel becomes even more expensive.

Wind energy experts are hopeful for the future of their industry. While infinite refinements and improvements are possible, the major technology developments that allow commercialization are complete. At some point, a "weather change" in the marketplace or a "killer application" somewhere will put several key companies or financial organizations in a position to profit, and wind energy conversion investors will take advantage of public interest, the political and economic climate, and emotional or marketing factors to po-sition wind energy technology (developed in a long lineage from the Chinese and the Persians to the present wind energy researchers and developers) for its next round of development (Spellman & Whiting, 2006).

Though wind energy production is generally considered unusually en-vironmentally clean, serious environmental issues do exist. For species protection, wind farm placement should be carefully studied. Wind farms put stresses on already fragmented and reduced wildlife habitat. Another serious factor is the avian mortality rate. Just as high-rise buildings, power lines, towers, antennas, and other man-made structures are passive killers of many birds, badly positioned wind farms put a heavy toll on bird popula-tions, especially upon migratory birds. The Altamont Pass wind farms (near San Francisco) are badly placed; since their construction in the 1980s, they have killed many golden eagles and other species as well. Golden eagles lock on to a prey animal and dive for it, totally blocking out the threat of the wind turbine. They can see the propellers under normal circumstances, but their instinctive prey focus is so strong that when they stoop over a kill, they see only their prey.

Six to 10 different companies, including U.S. Wind Power, Kenetech Windpower, and Green Mountain Energy, own the turbines at the Altamont Pass wind farms—over 7,000 of them. Another wind facility in Tehachapi Pass near Los Angeles poses little threat to bird populations.

In an interview with a reporter from the *San Francisco Chronicle*, conser-vationist Stan Moore states,

> It is estimated that 40 to 60 golden eagles are killed annually, plus 200 red-tailed hawks and smaller numbers of American kestrels, crows, burrowing owls and other birds. Those numbers are conservative. . . .
>
> I'm in favor of renewable energy when it is sited appropriately, but Altamont Pass is one of the worst places to put a wind farm on planet Earth, because it is adjacent to one of the densest breeding populations of golden eagles in the

world. It's a unique place for raptors because of the abundant food source in ground squirrels. . . .

Altamont Pass is not an appropriate place for wind turbines. What we have there is world-class golden-eagle habitat.

The California Energy Commission financed a 5-year study conducted by Dr. Grainger Hunt, a world authority on birds of prey who works with the Santa Cruz Predatory Bird Research Group. The study detected no population-level impacts for golden eagles in Altamont Pass; however, the local eagles could provide source population for all of California if the wind farm deaths were halted. Instead, the local eagles are an at-risk population: If other pressures disturbed the Altamont Pass golden eagle population—an outbreak of West Nile virus, for example—catastrophic population losses would occur because the wind turbines have removed much of the buffer population.

Because control guidelines are voluntary, not mandatory, the energy industry essentially polices itself on this issue. When the U.S. Fish and Wildlife Service (practicing what service officials themselves call "discretionary" law enforcement of service laws) chooses not to enforce the Migratory Bird Treaty Act and the Bald and Golden Eagle Protection Act, and when California officials fail to enforce their own decrees (a state designation of the golden eagle as a "fully protected species" and a "species of special concern"), the protections supposedly provided by federal and state laws become a farce (Pellissier, 2003).

References and Recommended Reading

Anthes, R. A. 1996. *Meteorology*, 7th ed. Upper Saddle River, NJ: Prentice Hall.

Anthes, R. A., Cahir, J. J., Fraizer, A. B., & Panofsky, H. A. 1984. *The Atmosphere*, 3rd ed. Columbus, OH: Charles E. Merrill Publishing Company.

Ingersoll, A. P. 1983. "The Atmosphere." *Scientific American*, 249(33): 162–174.

Lutgens, F. K., & Tarbuck, E. J. 1982. *The Atmosphere: An Introduction to Meteorology*. Englewood Cliffs, NJ: Prentice-Hall.

Miller, G. R., Jr. 2004. *Environmental Science*, 10th ed. Australia: Thompson-Brooks/ Cole.

Moran, J. M., Morgan, M. D., & Wiersma, J. H. 1986. *Introduction to Environmental Science*, 2nd ed. New York: W. H. Freeman & Company.

Pellissier, H. 2003. "Golden Eagle Eco-Atrocity at Altamont Pass," special to SF Gate: *San Francisco Chronicle*; Xcel Energy Ponnequin wind farm in northeastern Colorado. Accessed January 7, 2008, at http://telosnet.com/wind/.

Shipman, J. T., Adams, J. L., & Wilson, J. D. 1987. *An Introduction to Physical Science*, 5th ed. Lexington, MA: D. C. Heath & Company.

Spellman, F. R., & Whiting, N.E. 2006. *Environmental Science and Technology: Concepts and Applications*, 2nd ed. Rockville, MD: CRC Press.

11

Weather Forecasting Tools

Mackerel sky and mares' tails make tall ships carry low sails.

—Weather lore

It is best to read the weather forecast before praying for rain.

—Mark Twain

Butterfly Effect

IN CHAPTER 8, WE MENTIONED Edward Norton Lorenz's statement about a butterfly flapping its wings in Brazil causing a tornado in Texas. Lorenz's work with mathematical models demonstrated an important point about weather forecasting or predicting: Minute variations in the initial values of variables in his computer weather model would result in grossly divergent weather patterns (Palmer, 2008). This sensitive dependence on initial conditions came to be known as the butterfly effect (it also meant that weather predictions from more than a week out are generally fairly inaccurate; Lorenz, 1979). Thus, because of the so-called butterfly effect, weather forecasting beyond near-term predictions (within a few days) are bound to be inaccurate.

Notwithstanding the difficulty of making long-term weather forecast predictions, it is the short-term forecast that we are mostly concerned with. That is, in regard to pending weather conditions, what can we expect today? What can we expect tomorrow? What can we expect two or three days from now? These are the concerns that most of us relate to regarding future weather conditions. Fortunately, the science and practice of meteorology enable us to obtain fairly reliable short-term forecasts. In practicing meteorology, certain computer models, observations, common sense, and measuring instruments allow us to predict today's and tomorrow's weather with a high degree of accuracy.

Predicting Weather

To predict weather, meteorologists rely on a network of information supplied by weather stations and orbiting satellites. These sources provide data for large-scale mapping of the positions of large air masses circling the earth. Because air masses interact in a relatively predictable way, meteorologists are able to predict weather conditions and patterns in a relatively accurate manner.

Fronts, the zone along which the masses come into contact with each other, are the primary cause of most changes in weather. They occur when a large mass of cold air meets a large mass of warm air.

In regard to local weather, although fronts are important weather indicators, local weather can also be impacted or influenced by local geography. For example, coastal areas have more moderate temperatures than inland areas; hot air moves up slopes in hill areas during the day and down slopes at night; high-altitude areas are usually colder and receive more precipitation than low-altitude areas; air systems above urban areas are often warmer than the surrounding area, creating artificial low-pressure areas; and in coastal areas, cool air usually blows inland during the day and out to sea at night.

Meteorological Tools

(This section is based on information provided in EPA's 2011 *Basic Air Pollution Meteorology* information packet. Accessed July 15, 2011, at http://yosemite.epa.gov/oaqps/EOGtrain.nsf/.) In order to understand and predict weather events, it is imperative to understand the basic atmospheric processes that influence weather dynamics in the atmosphere. Measuring and recording meteorological variables provide the necessary information to forecast weather events. It is important to note that these same variables can be used to make qualitative and quantitative predictions of ambient air quality pollutant concentrations resulting from the release of pollutants.

Meteorological variables that influence weather events are transported and dispersed via wind speed and direction, turbulence intensity (i.e., atmospheric stability), temperature and temperature difference, solar radiation, mixing height, meteorological tool (instrument) accuracy, and quality assurance / quality control (QAQC) procedures.

In addition to those weather production parameters already discussed—barometric pressure, fronts, and local geographical features—the focus of this section will be to review the instrumentation needed to measure other meteorological variables most useful in weather forecasting practices, namely, wind speed and direction; ambient temperature and vertical temperature differences; solar radiation; and mixing height.

As mentioned earlier, many systems are available for measuring atmospheric parameters. Choosing the most appropriate sensors depends on the type of application for which the data is to be used. In addition to sensors, other equipment for signal conditioning, recording, and perhaps electronic data logging may be needed. Strict procedures for specifying, siting, and maintaining instruments need to be followed to ensure the collection of representative data.

Wind Speed

Although wind is a vector quantity and may be considered a primary variable in itself, it is more common to consider wind speed (the magnitude of the vector) and wind direction (the orientation of the vector) separately as variables. But let's consider the vector quantity relationship first. *Wind speed* or *velocity* is speed with direction. As stated, wind velocity is a vector quantity that refers to the rate at which an object or air mass changes its position. In regard to determining velocity, for example, it is important to know if you are traveling 50 mph due east or 50 mph due south. These same two speeds (but with different velocities) will take you to two very different final locations. Problems in physics generally involve velocities (represented by vectors, which are simply measured quantities that have both a direction and magnitude [or size]) because the direction of motion is typically an important piece of information.

Again, wind velocity is a vector quantity, and speed is a scalar quantity. Scalar quantities have magnitude but no direction. Velocity is defined as a change in distance in a given direction divided by a change in time. The Greek letter delta (Δ) is used to represent the concept of change. Thus, the equation

$$\text{Velocity} = \Delta x / \Delta t \qquad (11.1)$$

is read as velocity equals the change in x divided by the change in t. To determine how far an object has traveled from an initial position, after a set amount of time, traveling at constant velocity, we need to derive equation (11.2) from equation (11.1):

$$x = x_o + vt \tag{11.2}$$

where x = distance traveled, x_0 = initial position, t = a set amount of time, and v = constant velocity.

Example 11.1

Problem

You travel with an annual velocity of 60 mph, and you drive for 12 h on your second travel day, and your starting point (x_o) was 400 mi beyond where you were the day before; your total distance traveled at the end of the second day would be

Solution

$$x = (400 \text{ mi}) + (60 \text{ mph}) \times (12 \text{ h})$$
$$x = 1{,}120 \text{ mi}$$

In typical physics problems involving the velocity of an object, there are sometimes several velocities involved, and the velocity that results from the sum of all the velocities described (resultant velocity) must be determined. The resultant velocity is the sum of all the velocity vectors. A *vector* is a quantity that has direction as well as magnitude (size).

To gain understanding of vectors and vector math operations, imagine a speedboat crossing a river. If the speedboat were to point its bow straight towards the other side of the river, it would not reach the shore directly across from its starting point (see Figure 11.1). The river current influences the motion of the boat and carries it downstream. The speedboat may be moving with a velocity of 5 m/s directly across the river, yet the resultant velocity of the boat will be greater than 5 m/s and at an angle in the downstream direction. While the speedometer of the speedboat may read 5 m/s, its speed with respect to observers on the shore will be greater than 5 m/s. The resultant velocity of the motor boat can be determined by using vectors. The resultant velocity of the boat is the vector sum of the boat velocity and the river velocity. Because the speedboat heads across the river and because the current is always directed straight downstream the two vectors are at right angles to each other. The lengths of the sides of a right triangle are related by the Pythagorean theorem (see Figure 11.2), which states that in a right triangle the

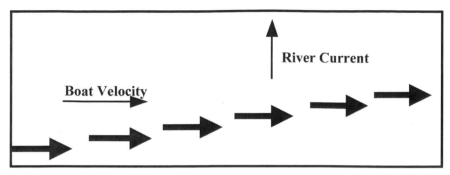

FIGURE 11.1
The motion of a speedboat with current.

square of the length of the long side (hypotenuse) is equal to the sum of the squares of the other two sides. Generally, the Pythagorean theorem is written as $a^2 + b^2 = c^2$.

In our example of the speedboat crossing the river, the two velocity vectors form a right triangle, so that the resultant velocity can be computed with the formula

$$v \text{ (resultant)} = (v_1^2 + v_2^2)^{1/2}$$

where v_1 is the velocity of the river, and v_2 is the velocity of the speedboat.

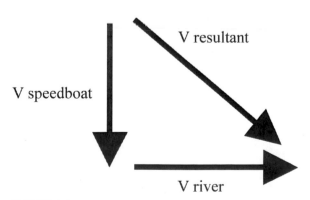

FIGURE 1.2
The lengths of the sides of a right triangle are related according to the Pythagorean theorem.

Example 11.2

Problem

Suppose that the river in our example is moving with a velocity of 4 m/s, north, and the speedboat is moving with a velocity of 5 m/s, east. What will be the resultant velocity of the speedboat (i.e., the velocity relative to observers on the shore)? The magnitude of the resultant can be found as follows:

Solution

$$(5.0 \text{ m/s})^2 + (4.0 \text{ m/s})^2 = R^2$$
$$25 \text{ m}^2/\text{s}^2 + 16 \text{ m}^2/\text{s}^2 = R^2$$
$$41 \text{ m}^2/\text{s}^2 = R^2$$
$$\sqrt{(41 \text{ m}^2/\text{s}^2)} = R$$
$$6.4 \text{ m/s} = R$$

Two main types of instruments that measure wind speed are the *rotating cup anemometer* and the *propeller anemometer* (rotating cup type described and depicted in chapter 2). Both types of anemometers consist of two sub-assemblies, the sensor and the transducer. The sensor is the device that rotates by the force of the wind. The transducer is the device that generates the signal suitable for recording. A complete instrument package may also include electronics to capture and record the electronic signals generated by the transducer. For instance, the signal may need to be conditioned so that a reportable quantity can be derived from the signal. This is accomplished with a signal conditioner. Finally, the conditioned signal needs to be displayed and/or recorded for it to be usable. Data recorders and loggers accomplish this task.

Note: We already discussed anemometers and their operation in chapter 2 of this text. The reader may want to review this information before proceeding to the following discussion on wind speed transducers.

Wind Speed Transducers

There are several mechanisms that can be used to convert the rate of the cup or propeller rotations to an electrical signal suitable for recording and/or processing. The selection of the transducer is determined by the nature of the monitoring program—how responsive the instrument needs to be and what type of data readout or recoding is needed. The four most commonly used types of transducers are the DC generator, the AC generator, the electrical contact, and the interrupted light beam. Many *DC* and *AC generator* types of transducers in common use have limitations in terms of achieving low thresholds and quick response times. It is important to use instruments, such as those anemometers that employ miniaturized DC generators, that have low starting thresholds. The AC generator transducers eliminate the brush friction, but the signal conditioning circuitry must be carefully designed to avoid spurious oscillations in the output signal that may be produced at lower wind speeds. *Electric-contact* transducers are used to measure the total passage of the wind (wind-run) instead of instantaneous wind speeds and may be used to determine the average wind speed over a given time increment. These devices are typically not appropriate for use in air pollutant dispersion studies. The *interrupted light beam* (light chopping) transducer is frequently used in air quality applications because it exhibits less friction and it is more responsive to lower wind speeds. This type of transducer uses either a slotted shaft or a slotted disk, a photo emitter, and a photo detector. The cup or propeller assembly rotates the slotted shaft or disk, creating a pulse each time the light passes through a slot and falls on the photo detector.

The frequency output from an AC generator or a light chopping transducer may be transmitted through a signal conditioner and converted to an analog signal for various recording devices, such as a continuous strip chart or a multipoint recorder, or through an analog-to-digital (A/D) converter to a microprocessor type of digital recorder. Several modern data-loggers can accept the frequency type signal directly, eliminating the need for additional signal conditioning. The recording and processing of the data need to be considered in designing a monitoring program.

Wind Direction

Wind direction is generally defined as the orientation of the wind vector in the horizontal. Wind direction for meteorological purposes is defined as the direction from which the wind is blowing and is measured in degrees clockwise from true north. For example, a westerly wind is blowing from the west, 270° from north. A north wind is blowing from a direction of 360°. Wind direction determines the transport direction of an emitted plume.

Wind Vanes

The most common instrument for measuring wind direction is the wind vane. Wind vanes point in the direction from which the wind is flowing. Wind vanes come in many different shapes and sizes: some with two plates joined at their forward edges and spread out at an angle (splayed vanes) and others with a single flat plate or perhaps a vertical airfoil. Vanes are commonly con-structed from stainless steel, aluminum, or plastic. As with anemometers, care should be taken in selecting a sensor that has a proper balance of durability and sensitivity for a particular application. An example of a wind vane type is depicted and described in chapter 2.

The horizontal (azimuth) and vertical (elevation) components of the wind direction can be measured with a bi-directional wind vane (bivane). The bi-vane generally consists of either an annular fin or two flat fins perpendicular to each other, counterbalanced and mounted on a gimbal so that the unit can rotate freely both horizontally and vertically.

Fixed-Mount Propeller Anemometers

Another method of obtaining the horizontal and/or vertical wind direction is through the use of orthogonal fixed-mount propeller anemometers (men-tioned earlier). The horizontal wind direction can be determined computa-tionally from the orthogonal wind speed components. Vertical velocity can also be measured by adding a third propeller mounted vertically. This device is often referred to as a UVW anemometer.

Wind Direction Transducers

Many kinds of simple commutator-type transducers utilize brush contacts to divide the wind direction into 8 or 16 compass point sectors. However, transducers that provide at least 10° resolution (36 compass point sectors) in wind direction measurements are preferred for use in air quality applications.

A fairly common transducer for air quality modeling applications is a potentiometer. The voltage across the potentiometer varies directly with the wind direction. A potentiometer is a variable resistor. When the wind direc-tion changes, the shaft of the wind vane moves and changes the resistance across the potentiometer. This change is directly related to wind direction.

Siting and Exposure of Wind Measuring Instruments

Proper instrument siting is a key to obtaining representative meteoro-logical data for use in air pollution studies. Instruments need to be located in areas away from obstructions that can influence the measurements. Second-ary considerations such as accessibility and security should not be allowed to compromise data quality.

The standard exposure height of wind instruments over level, open ter-rain is 10 meters above the ground. Open terrain is defined as an area where the distance between the instrument and obstruction (trees, buildings, etc.) is at least 10 times the height of the obstruction. In some cases where emission releases occur substantially above 10 meters, additional wind measurements may be needed at higher elevations. Appropriate measurement heights would be established on a case-by-case basis depending upon the application. Mounting wind instruments on an open-lattice tower is recommended when possible. Tower-mounted wind instruments should be placed on the top of the tower, or, if mounted on the side of the tower, instruments should be placed on booms at a distance of at least twice the diameter/diagonal of the tower extending outward toward the prevailing wind direction.

Temperature and Temperature Difference

Both ambient air temperature at a single level (typically 1.5 m to 2 m above ground) and temperature difference between two levels (typically 2 m and 10 m) are useful in weather forecasting and air pollution studies. These temperature measurements are useful in calculations of plume rise and can be used in determining atmospheric stability.

There are three main classes of temperature sensors based on (1) thermal expansion; (2) resistance change; and (3) thermoelectric properties of various substances as a function of temperature. The alcohol and mercury liquid-in-glass bulb thermometers are common examples of thermal expansion sensors. However, these are of limited value in on-site or remote monitoring networks because they lack the means for automated data recording.

A common type of sensor for on-site meteorological measurement programs is the resistance temperature detector (RTD). The RTD operates on the basis of the resistance changes of certain metals, usually platinum or copper, as a function of temperature. These two metals are the most commonly used because they show a fairly linear increase of resistance with rising temperature. A second type of resistance change thermometer is the thermistor, which is made from a mixture of metallic oxides fused together. The thermistor generally gives a larger resistance change with temperature than the RTD. Because the relation between resistance and temperature for a thermistor is nonlinear, systems generally are designed to use a combination of two or more thermistors and fixed resistors to produce a nearly linear response over a specific temperature range.

Did You Know?

Thermistors make superior thermometers because they are small, stable, long lasting, and accurate.

Thermoelectric sensors work on the principle of a temperature dependent electrical current flow between two dissimilar metals. Such sensors, called thermocouples, have some special handling requirements of installation in order to avoid induction currents from nearby AC sources, which can cause errors in measurement. Thermocouples are also susceptible to spurious voltages caused by moisture. For these reasons, their usefulness for routine field measurements is limited.

Temperature Difference

The basic sensor requirements for measuring vertical temperature difference are essentially the same as for single ambient temperature measurement. However, matched sensors and careful calibration are required to achieve the desired accuracy of measurement.

Siting and Exposure of Instruments to Measure Temperature and Temperature Difference

Ambient air (surface) temperature should be measured at a height of 2 meters. The standard height for temperature difference measurements is 2 and 10 meters. If elevated emissions levels are a concern, it may be appropriate to make additional temperature measurements at height elevations. These elevations would be determined on a case-by-case basis depending upon the application. The temperature sensor should be located over an open, level, and well-ventilated area of a least 9 meters in diameter. Additionally, temperature sensors should be located at a distance of at least four times the height of any obstruction and at least 30 meters from larger paved areas. The surface under the temperature sensor should be covered by a natural earth surface or short grass and located away from low-lying areas that hold standing water. Instruments should be shielded to protect them from thermal radiation and should also be well ventilated using aspirated shields.

Radiation

Solar radiation is related to the stability of the atmosphere. Cloud cover and ceiling height (height of the base of the cloud deck that obscures at least half the sky) data, taken routinely at National Weather Service (NWS) stations, provide an indirect estimation of radiation effects and are used in conjunction with wind speed to derive an atmospheric stability category. If representative information is not available from routine NWS observations, it may be appropriate to measure solar radiation for use in determining atmospheric stability.

An instrument that is used most frequently to measure solar radiation is a pyranometer. The *pyranometer* measures direct and diffuse radiation in watts per meter square on a horizontal surface. In this instrument is a small flat disk with sectors of the disk painted alternating black and white. When exposed to solar radiation the black sectors become warmer than the white areas. This temperature difference is detected electronically. An electrical voltage proportional to the incoming solar radiation energy is produced. A standard optical glass dome is installed over the disk that is transparent to wavelengths from about 280 to 2,800 nanometers. Some pyranometers use a silicon glass dome in order to measure radiation in different spectral intervals.

Another type of sensor is the *net radiometer*, which is an instrument that measures the net difference between the upward and downward components of solar and terrestrial radiation. More specifically, the net radiometer is a four-component device that serves to measure the four separate components of the surface radiation balance: global solar radiation, reflected solar radiation, infrared emitted by the sky, and infrared emitted by the ground surface. The primary application of a net radiometer would be to determine the daytime and nighttime radiation balance as an indicator of stability. However, nighttime stability categories commonly used in air pollution studies are based solely on wind speed and cloud-cover conditions.

Siting and Exposure of Radiation Measuring Instruments

Pyranometers used for measuring incoming (solar) radiation should be located in open areas with an unrestricted view of the sky in all directions during all seasons. They should be located to avoid obstructions casting a shadow on the sensor at any time. Also, sitting the instrument near light-colored walls and artificial sources of radiation should be avoided. Sensor height is not critical for pyranometers. A tall platform or rooftop is a desirable location.

Net radiometers should be mounted about 1 meter above the ground. The ground cover under a net radiometer should be representative of the general site area. Net radiometers should also be located to avoid obstructions to the field of view both upward and downward.

Mixing Height

The vertical depth of the atmosphere where mixing takes place is called the mixing layer (i.e., the troposphere—*tropos* from the Greek for turning or mixing). The top of the mixing layer is referred to as the mixing height. The mixing height determines the vertical extent of the dispersion process for air pollutants that are released below the mixing height. The mixing height is an important variable in air quality studies as it restricts vertical dispersion of pollutants. Although mixing heights are not typically measured directly, they can be approximated from routine meteorological measures.

Morning and afternoon mixing heights are typically estimated from the vertical temperature profiles observed at selected National Weather Service stations taken at 1200 Greenwich Mean Time (GMT) and surface temperature measurements. Vertical temperature profiles are measured with radiosondes, instruments transported aloft by lighter-than-air balloons (i.e., balloons typically filled with either hydrogen or helium). For air quality modeling studies used in regulatory applications, hourly mixing heights are estimated from the twice-daily mixing height values, sunrise and sunset times, and hourly stability categories by the meteorological preprocess computer program developed for use with EPA regulatory models.

The Doppler SODAR (an acronym for **SO**und **D**etection **A**nd **R**anging) systems are gaining recognition as effective tools for remote measurement of meteorological variables at heights up to several hundred meters above the surface. A SODAR transmits a strong acoustic pulse into the atmosphere and listens for that portion of the pulse that is scattered and returned. There has been an increased interest in using SODARs to develop the meteorological databases required as input to weather and dispersion models. SODAR returns can also be analyzed to estimate the mixing height.

System Performance

Monitoring the proper meteorological variables that are representative of atmospheric conditions at a specific location is crucial in a weather monitoring program. It is equally important that the performance of the monitoring system be adequate to produce representative data. The accuracy and response characteristics of meteorological monitoring systems are important factors in defining system performance.

System Accuracy

Accuracy is the degree of agreement between measured values and the accepted reference value. The investigator must carefully design his sampling program and use certain statistical tools to evaluate his data before making any inferences from the data. In a target shooting analogy, accuracy can be equated to how close does a target shooter come to the bull's-eye? *Precision*, on the other hand, is the reproducibility of replicate analyses of the same sample (mass or concentration). For example, how close to each other is a target shooter able to place a set of shots anywhere on the target? *Bias* is the error introduced into sampling that causes estimates of parameters to be inaccurate. More specifically, bias is the difference between the average measured mass or concentration and a reference mass or concentration, expressed as a fraction of the reference mass or concentration. For example, how far from the bull's-eye is the target shooter able to place a cluster of shots?

In regard to meteorological measuring devices, system accuracy is the amount by which a measured variable deviates from a value accepted as true or standard. Accuracy can be thought of in terms of individual component accuracy or overall system accuracy. For example, the overall accuracy of a wind speed measurement system includes the individual component accuracies of the cup or propeller anemometer, electronic circuitry such as a signal conditioner, and data recorder.

The accuracies recommended for on-site meteorological monitoring systems producing data to be used in regulatory modeling applications are stated in terms of overall system accuracies, since it is the data from the measurement system that are used in air quality analyses. Recommended measurement resolutions, that is, the smallest increments that can be distinguished, are also provided in Table 11.1. These resolutions are considered necessary to maintain the recommended accuracies.

The accuracy specifications and resolutions provided in Table 11.1 are applicable to the primary measurement system, which is recommended to be a microprocessor-based digital system. For analog systems used as back-up the recommended accuracy limits in Table 11.1 may be increased by 50%.

TABLE 11.1
Recommended System Accuracies and Resolutions

Meteorological Variable	System Accuracy	Measurement Resolution
Wind speed	± (0.2 m/s + 5% of observed)	0.1 m/s
Wind direction	± 5°	1°
Ambient temperature	± 0.5°C	0.1°C
Vertical temperature difference	± 0.1°C	0.02°C
Radiation	± 5% of observed or 25 W/m2*	10 W/m2
Time	± 5 min	

*Whichever is greater

Source: US EPA, 1987. *On-Site Meteorological Program Guidance for Regulatory Modeling.*

Resolutions for such analog systems should be adequate to maintain the recommended accuracies.

Response Characteristics of On-Site Meteorological Sensors

Response characteristics help define how quickly an instrument will respond to changing meteorological variables. Certain response characteristic of meteorological sensors proposed for on-site monitoring programs must be known to ensure that data being collected are appropriate for the intended application.

The following definitions apply for terms commonly associated with instrument response characteristics and inherent properties of meteorological sensors:

Calm. Any average wind speed below the starting threshold of the wind speed or direction sensor, whichever is greater.

Damping ratio. The motion of a wind vane is a damped oscillation, and the ratio in which the amplitude of successive swings decreases is independent of wind speed. The damping ratio is the ratio of actual damping to critical damping, which is a measure of a vane's mechanical resistance to investment.

Delay distance. The length of a column of air that passes a wind vane such that the vane will respond to 50% of a sudden angular change to wind direction.

Distance constant. The distance constant of a sensor is the length of fluid flow past the sensor required to cause it to respond to a given change in wind speed. Distance constant is a characteristic of cup and propeller (rotational).

Range. This is a general term that usually identifies the limits of operations of a sensor, most often within which the accuracy is specified.

Threshold (stating speed). The wind speed at which an anemometer or vane first starts to perform within it specifications.

Time constant. The period of time that is required for a sensor to respond to a given change in the parameter the sensor in measuring.

Table 11.2 lists recommended sensor response characteristics for use in regulatory modeling applications.

Quality Assurance and Quality Control

Quality assurance (QA) applied to meteorological monitoring consists of both "the system of activities to produce a quality product" (traditional quality control) and "the system of activities to produce assurance that the qual-

TABLE 11.2
Recommended Response Characteristics for Meteorological Sensors

Meteorological Variables	Sensor Specification(s)
Wind speed	Starting Speed ≤ 0.5 m/s
	Distance constant ≤ 5 m
Wind direction	Starting speed ≤ 0.5 m/s
	Damping ratio 0.4 to 0.7
	Delay distance ≤ 5 m
Temperature	Time constant ≤ 1 min
Temperature difference	Time constant ≤ 1 min
Radiation	Time constant – 5 sec
	Operating temperature range –20°C to +40°C at specified accuracy

Source: US EPA, 1987.

ity control system is performing adequately" (traditional quality assurance) (Finkelstein et al., 1983). The first of these quality control (QC) functions consists of those activities performed by equipment operators directly on the instruments, that is, preventive maintenance, calibrations, and so on. The purpose of the second set of activities is to manage the quality of the data and administer corrective actions as necessary to ensure that the data quality requirements are met. Formal plans for quality assurance must be presented in a document called the AQ plan. This document lists all necessary quality-related procedures and the frequency with which they should be performed. It is imperative that a QA plan be developed and followed to ensure that representative data of good quality are obtained. Specific information to be included in a QA plan is described below.

Project personnel responsibilities. Responsibilities of personnel performing tasks that affect data quality.

Data reporting procedures. Brief description of how data are produced, delineating functions performed during each step of the data processing sequence.

Data validation procedures. Detailed listing of criteria to be applied to data for testing their validity, how the validation process is to be carried out. And the treatment of data found to be questionable or invalid.

Audit procedures. Detailed description of what audits are to be performed, how often they are to be performed, and an audit procedure (referencing document procedures whenever possible). Also, description of internal and external systems audits including site inspections by supervisory personnel or others.

Calibration procedures. Detailed description of calibration techniques and frequency for calibrating each of the sensors or instruments being used. Both full calibrations and zero and span checks should be defined.

Preventive maintenance schedule. Detailed listing of specific preventive maintenance functions and the frequency at which they should be performed. Includes not only routine equipment inspection and wearable parts replacement but also functional tests to be performed on equipment.

Quality reports. The schedule and content of reports submitted to management describing the status of the quality assurance program. The quality assurance program includes the implementation of all functions specified in the QA plan. This implementation involves personnel at all levels of the organization. Technicians who operate equipment must perform preventive maintenance and QC checks on the measurements system for which they are responsible. They must perform calibrations and, when required, participate in the internal audits of stations run by other technicians. Their immediate supervisors should check to see that all specified QA tasks are performed, and they should review logs and control charts to ensure that potential problems are corrected before significant data loss occurs.

References and Recommended Reading

Finkelstein, P. L., et al. 1983. *Quality Assurance Handbook for Air Pollution Measurement Systems, Vol. IV: Meteorological Measurements*. Research Triangle Park, NC.

Lorenz, E. N. 1979. *Butterfly Effect*. Recorded from Lorenz's address at the annual meeting of the American Association for the Advancement of Science.

Palmer, T. 2008. "Edward Norton Lorenz." *Physics Today* 61(9): 81–82.

U.S. EPA. 1987. *On-Site Meteorological Program Guidance for Regulatory Modeling Applications*. Accessed July 25, 2011, at http://Nepis.epa.gov/exe/ZyN.exe.

Glossary

Note: This glossary was compiled from information by the author and several sources cited by NOAA (2011) in www.erh.noaa.gov/er/box/glossary.htm.

abiotic: The nonliving part of the physical environment (for example, light, temperature, and soil structure).

ablation: The process of being removed. Snow ablation usually refers to removal by melting.

absolute humidity: The density of water vapor. It is the mass of the water vapor divided by the volume that it occupies.

absorption: (1) Movement of a chemical into a plant, animal, or soil; (2) any process by which one substance penetrates the interior of another substance. In chemical spill cleanup, this process applies to the uptake of chemicals by capillaries within certain sorbent materials.

absorption units: Devices or units designed to transfer the contaminant from a gas phase to a liquid phase.

accessory clouds: Clouds that are dependent on a larger cloud system for development and continuances. Accessory clouds associated with the thunderstorm include roll, shelf, mammatus, and wall clouds.

accidental spills: The unintended release of chemicals and hazardous compounds or materials into the environment.

accretion: Growth of precipitation particles by collision of ice crystals with supercooled liquid droplets that freeze on impact.

acid: A hydrogen-containing corrosive compound that reacts with water to produce hydrogen ions; a proton donor; a liquid compound with a pH less than or equal to 2.

acid rain: Precipitation made more acidic from falling through air pollutants (primarily sulfur dioxide) and dissolving them.

acidic deposition: See *acid rain.*

additive data: A group of coded remarks in a weather observation that includes pressure tendency, amount of precipitation, and maximum/minimum temperature during specified periods of time.

adiabatic: Without loss or gain of heat. When air rises, air pressure decreases and expands adiabatically in the atmosphere; since the air can neither gain nor lose heat its temperature falls as it expands to fill a larger volume.

adiabatic lapse rate: The temperature profile or lapse rate, used as a basis for comparison for actual temperature profiles (from ground level) and hence for predictions of stack gas dispersion characteristics.

adsorption: (1) The process by which one substance is attracted to and adheres to the surface of another substance without actually penetrating its internal structure; (2) process by which a substance is held (bound) to the surface of a soil particle or mineral in such a way that the substance is only available slowly.

adsorption site density: The concentration of sorptive surface available from the mineral and organic contents of soils. An increase in adsorption sites indicates an increase in the ability of the soils to immobilize hydrocarbon compounds in the soil matrix.

advection fog: A type of fog that results from the advection of moist air over a cold surface and the cooling of the air to its dew point that follows; this type of fog is most common in coastal regions.

advective wind: The horizontal air movements resulting from temperature gradients that give rise to density gradients and subsequently pressure gradients.

advisory: Advisories are issued for weather situations that cause significant inconvenience but do not meet warning criteria and, if causation is not exercised, could lead to life-threatening situations. Advisories are issued for significant events that are occurring, are imminent, or have a very high probability of occurrence.

aerobic: Living in the air. Opposite of *anaerobic.*

aerobic processes: Many bio-technology production and effluent treatment processes are dependent on microorganisms that require oxygen for their metabolism. For example, water in an aerobic stream contains dissolved oxygen. Therefore, organisms using this can oxidize organic wastes to simple compounds.

aerosol: Particles of matter, solid or liquid, larger than a molecule but small enough to remain suspended in the atmosphere (up to 100 μm diameter). Natural origins include salt particles from sea spray and clay particles as a result of weathering of rocks. Aerosols can also originate as a result of man's activities and in this case are often considered pollutants.

aerovane: Aerovanes are commonly used at many weather stations and airports to measure both wind direction and speed. They are similar to wind vanes and cup anemometers except for three-bladed propellers attached to the end of the vane.

AFOS: The Automation of Field Operations and Services; AFOS is the computer system that links National Weather Service offices and other computer networks, such as the NOAA Weather Wire, to transmit weather information.

AGL: Above ground level.

agricultural sources: Both organic and inorganic contaminants usually produced by pesticide, fertilizers, and animal wastes, all of which enter water bodies via runoff and groundwater absorption in areas of agricultural activity.

air: The mixture of gases that constitutes the earth's atmosphere.

air currents: Created by air moving upward and downward.

air mass: A large body of air with particular characteristics of temperature and humidity. An air mass forms when air rests over an area long enough to pick up the conditions of that area.

airmass thunderstorm: Generally, a thunderstorm not associated with a front or other type of synoptic-scale forcing mechanism. Air mass thunderstorms typically are associated with warm, humid air in the summer months; they develop during the afternoon in response to insolation and dissipate rather quickly after sunset.

air parcel: An imaginary small body of air that is used to explain the behavior of air. A parcel is large enough to contain a very great number of molecules but small enough so that the properties assigned to it are approximately uniform throughout.

air pollutants: Generally include sulfur dioxide, hydrogen sulfide, hydrocarbons, carbon monoxide, ozone, and atmospheric nitrogen but can include any gaseous substance that contaminates air.

air pollution: Contamination of the atmosphere with any material that can cause damage to life or property.

air pressure: Atmospheric pressure. Air pressure is the force exerted on a surface by the weight of the air above it. The internationally recognized unit for measuring this pressure is the kilopascal.

air stripping: A mass transfer process in which a substance in solution in water is transferred to solution in a gas.

airborne contaminants: Any contaminant capable of dispersion in air and/ or capable of being carried by air to other locations.

airborne particulate matter: Fine solids or liquid droplets suspended and carried in the air.

airstream: A significant body of air flowing in the same general circulation.

albedo: The fraction of received radiation reflected by a surface. Snow covered areas have a high albedo (0.9 or 90%) due to their white color.

Alberta Clipper: A small, fast-moving low-pressure system that forms in western Canada and travels southeastward into the United States. These storms, which generally bring little precipitation, usually precede an Arctic air mass.

algae: A large and diverse assemblage of eukaryotic organisms that lack roots, stems, and leaves but have chlorophyll and other pigments for carrying out oxygen-producing photosynthesis.

aliphatic hydrocarbon: Compound comprised of straight-chain molecules as opposed to a ring structure.

alkalinity: (1) The concentration of hydroxide ions; (2) the capacity of water to neutralize acids because of the bicarbonate, carbonate, or hydroxide content. Usually expressed in milligrams per liter of calcium carbonate equivalent.

alkanes: A class of hydrocarbons (gas, solid, or liquid depending upon carbon content). Its solids (paraffins) are a major constituent of natural gas and petroleum. Alkanes are usually gases at room temperature (methane) when containing less than five carbon atoms per molecule.

alkenes: A class of hydrocarbons (also called olefins); sometimes gases at room temperature but usually liquids; common in petroleum products. Generally more toxic than alkanes, less toxic than aromatics.

alkynes: A class of hydrocarbons (formerly known as acetylenes). Unsaturated compounds characterized by one or more triple bonds between adjacent carbon atoms. Lighter alkenes, such as ethyne, are gases; heavier ones are liquids or solids.

altimeter: An active instrument used to measure the altitude of an object above a fixed level.

altimeter setting: That pressure value to which an aircraft altimeter scale is set so that it will indicate the altitude above mean sea level of an aircraft on the ground at the location for which the value was determined.

altitude: Height expressed as the distance above a reference point, which is normally sea level or ground level.

altocumulus: Mid-altitude clouds with a cumuliform shape.

altostratus: Mid-altitude clouds with a flat, sheetlike shape.

anabatic: Wind flowing up an incline, such as up a hillside; upslope wind.

anaerobic: Not requiring oxygen.

anaerobic process: Any process (usually chemical or biological) carried out without the presence of air or oxygen (for example, in a heavily polluted watercourse with no dissolved oxygen present).

anafront: A front at which the warm air is ascending the frontal surface up to high altitudes.

analysis: The separation of an intellectual or substantial whole into its constituent parts for individual study.

anemometer: An instrument that measures wind speed.

aneroid barometer: An instrument built around a metal structure that bends with changing air pressure. These changes are recorded on a pointer that moves back and forth across a printed scale.

angular momentum: The energy of motion of a spinning body or mass of air or water.

angular velocity: The rate at which a spinning body rotates.

anomaly: The deviation of (usually) temperature or precipitation in a given region over a specified period from the normal value for the same region.

anthropogenic sources: Generated by human activity.

anticyclone: High-atmosphere areas characterized by clear weather and the absence of rain and violent winds.

anticyclonic: Describes the movement of air around a high pressure and rotation about the local vertical opposite the earth's rotation. This is clockwise in the northern hemisphere.

anvil cloud: The flat, spreading top of a Cb (cumulonimbus), often shaped like an anvil. Thunderstorm anvils may spread hundreds of miles downwind from the thunderstorm itself and sometimes may spread upwind.

anvil crawler: A lightning discharge occurring within the anvil of a thunderstorm, characterized by one or more channels that appear to crawl along the underside of the anvil. They typically appear during the weakening or dissipating stage of the parent thunderstorm or during an active MCS (mesoscale convective system).

anvil dome: A large overshooting top or penetrating top.

anvil rollover: A circular or semicircular lip of clouds along the underside of the upwind part of a back-sheared anvil, indicating rapid expansion of the anvil.

anvil zits: Frequent (often continuous or nearly continuous), localized lightning discharges occurring from within a thunderstorm anvil.

arctic air: A mass of very cold, dry air that usually originates over the Arctic Ocean north of Canada and Alaska.

arctic high: A very cold high pressure that originates over the Arctic Ocean.

arcus: A low, horizontal cloud formation associated with the leading edge of thunderstorm outflow (i.e., the gust front). Roll clouds and shelf clouds both are types of arcus clouds.

aridity: A general term used to describe areas suffering from lack of rain, or drought. More specifically, a condition in which evaporation exceeds precipitation.

aromatic hydrocarbons: Class of hydrocarbons considered to be the most immediately toxic; found in oil and petroleum products; soluble in water. Antonym: aliphatic.

ASOS: Automated Surface Observing System. This system observes sky conditions, temperature and dew point, wind direction and speed, barometric pressure, and precipitation.

atmosphere: The layer of air surrounding the earth's surface.

atmospheric pressure: Also called air pressure or barometric pressure. The pressure asserted by the mass of the column of air directly above any specific point.

atmospheric stability: An indication of how easily a parcel of air is lifted. If the air is very stable it is difficult to make the parcel rise. If the air is very unstable the parcel may rise on its own once started.

atom: A basic unit of physical matter indivisible by chemical means, the fundamental building block of chemical elements; composed of a nucleus of protons and neutrons surrounded by electrons.

atomic number: Number of protons in the nucleus of an atom. Each chemical element has been assigned a number in a complete series from 1 to 100+.

atomic orbitals / electron shells: The region around the nucleus of an atom in which an electron is most likely to be found.

atomic weight: The mass of an element relative to its atoms.

auger: A tool used to bore holes in soil to capture a sample.

aurora borealis: Also known as the northern lights—the luminous radiation emission from the upper atmosphere over middle and high latitudes, centered around the earth's magnetic poles. These silent fireworks are often seen on clear winter nights in a variety of shapes and colors.

automated weather station: An unmanned station with various sensors that measure weather elements such as temperature, wind, and pressure and transmit these reading for use by meteorologists.

automatic samplers: Devices that automatically take samples from a waste stream.

autotrophic: An organism that can synthesize organic molecules needed for growth from inorganic compounds using light or another source of energy.

autotrophs: See *autotrophic*.

avalanche: A large mass of rapidly moving snow down a steep mountain slope.

AVHRR: Advanced Very High Resolution Radiometer. Main sensor on U.S. polar-orbiting satellites.

AVN: Aviation model generated every 12 hours by NCEP.

Avogadro's number: The number of carbon atoms in 12 grams of the carbon-12 isotope (6.022045×1023). The relative atomic mass of any element, expressed in grams, contains this number of atoms.

AWIPS: Advanced Weather Information Processing System. New NWS computer system integrating graphics, satellite, and radar imagery. The successor to AFOS.

bacilli (pl.), **bacillus** (sing): Members of a group of rodlike bacteria that occur everywhere in soil and air. Some are responsible for diseases such as anthrax or for causing food spoilage.

back door cold front: A front that moves east to west in direction rather than the normal west to east movement. For instance, one that enters southern New England from the Gulf of Maine.

back-building thunderstorm: A thunderstorm in which new development takes place on the upwind side (usually the west or southwest side), such that the storm seems to remain stationary or propagate in a backward direction.

back-sheared anvil: A thunderstorm anvil that spreads upwind, against the flow aloft. A back-sheared anvil often implies a very strong updraft and a high severe weather potential.

backing wind: Wind that shifts in a counterclockwise direction with time at a given location (e.g., from southerly to southeasterly) or changes direction in a counterclockwise sense with height (e.g., westerly at the surface but becoming more southerly aloft). Backing winds with height are indicative of cold air advection (CAA). The opposite of veering winds.

bacteria: One-celled microorganisms.

bacteriophage: A virus that infects bacteria; often called a *phage*.

baghouse filter: A closely woven bag for removing dust from dust-laden gas streams. The fabric allows passage of the gas with retention of the dust.

ball lightning: A relatively rare form of lightning, generally consisting of an orange or reddish ball of the order of a few cm to 30 cm in diameter and of moderate luminosity, which may move up to 1 m/s horizontally with a lifetime of a second or two.

barber pole: A thunderstorm updraft with a visual appearance including cloud striations that are curved in a manner similar to the stripes of a barber pole. The structure typically is most pronounced on the leading edge of the updraft, while drier air from the rear-flank downdraft often erodes the clouds on the trailing side of the updraft.

baroclinic zone: A region in which a temperature gradient exists on a constant pressure surface. Baroclinic zones are favored areas for strengthening and weakening systems.

barogram: The graphic record of pressure produced by a barograph.

barograph: An instrument that provides a continuous record of atmospheric pressure.

barometer: An instrument for measuring atmospheric pressure.

barometric pressure: The actual pressure value indicated by a pressure sensor.

barometric tendency: The amount and direction of change in barometer readings over a three-hour period.

barotropic system: A weather system in which temperature and pressure surfaces are coincident—that is, temperature is uniform (no temperature gradient) on a constant pressure surface. Barotropic systems are characterized by a lack of wind shear and thus are generally unfavorable areas for severe thunderstorm development.

base: A substance that when dissolved in water generates hydroxide (OH–) ions or is capable of reacting with an acid to form a salt.

bear's cage: A region of storm-scale rotation, in a thunderstorm, that is wrapped in heavy precipitation. This area often coincides with a radar hook echo and/or mesocyclone, especially one associated with an HP storm. The term reflects the danger involved in observing such an area visually, which must be done at close range in low visibility.

Beaufort scale: A scale that indicates the wind speed using the effect wind has on certain familiar objects.

beaver('s) tail: A particular type of inflow band with a relatively broad, flat appearance suggestive of a beaver's tail. It is attached to a supercell's general updraft and is oriented roughly parallel to the pseudo-warm front— that is, usually east to west or southeast to northwest.

benthic (benthos): The term originates from the Greek word for bottom and broadly includes aquatic organisms living on the bottom or on submerged vegetation.

best available technology (BAT): Essentially a refinement of best practicable means whereby a greater degree of control over emissions to land, air, and water may be exercised using currently available technology.

binomial system of nomenclature: A system used to classify organisms; organisms are generally described by a two-word scientific name, the *genus* and *species*.

biochemical oxygen demand (BOD): The amount of oxygen required by bacteria to stabilize decomposable organic matter under aerobic conditions.

biodegradable: A material capable of being broken down, usually by microorganisms, into basic elements.

biodegradation: The ability of natural decay processes to break down manmade and natural compounds to their constituent elements and compounds, for assimilation in, and by, the biological renewal cycles (e.g., as wood is decomposed to carbon dioxide and water).

biogeochemical cycles: *Bio* refers to living organisms and *geo* to water, air, rocks, or solids. *Chemical* is concerned with the chemical composition of the earth. Biogeochemical cycles are driven by energy, directly or indirectly, from the sun.

biological oxygen demand (BOD): The amount of dissolved oxygen taken up by microorganisms in a sample of water.

biology: The science of life.

biosphere: The region of the earth and its atmosphere in which life exists, an envelope extending from up to 6,000 meters above to 10,000 meters below sea level that embraces all life from Alpine plant life to the ocean deeps.

biota: The animal and plant life of a particular region considered as a total ecological entity.

biotic: Pertaining to life or specific life conditions.

black ice: Thin, new ice that forms on fresh water or dew covered surfaces; it is common on roadways during the fall and early winter and appears "black" because of its transparency.

blizzard: Includes winter storm conditions of sustained winds or frequent gusts of 35 mph or more that cause major blowing and drifting of snow, reducing visibility to less than one-quarter mile for three or more hours. Extremely cold temperatures often are associated with dangerous blizzard conditions.

blizzard warning: Issued when blizzard conditions are expected or are occurring.

blocking high: A high-pressure area (anticyclone), often aloft, that remains nearly stationary or moves slowly compared to west-to-east motion. It blocks the eastward movement of low-pressure areas (cyclones) at its latitude.

blowing dust: Dust that is raised by the wind to moderate heights above the ground to a degree that horizontal visibility decreases to less than seven miles. Visibilities of one-eighth mile or less over a widespread area are criteria for a blowing dust advisory.

blowing sand: Sand particles picked up from the surface of the earth by the wind to moderate heights above the ground, reducing the reported horizontal visibility to less than seven statute miles.

blowing snow: Wind driven snow that reduces visibility to six miles or less causing significant drifting. Blowing snow may be snow that is falling and/or loose snow on the ground picked up by the wind.

blowing spray: Water droplets torn by the wind from a body of water, generally from the crests of waves, and carried up into the air in such quantities that they reduce the reported horizontal visibility to less than seven statute miles.

blustery: Descriptive term for gusty winds that accompany cold weather.

boiling point: The temperature at which a substance changes from a liquid to a gas.

bomb cyclone: An extratropical area of low pressure in which the central pressure drops at least 24 millibars in 24 hours.

boundary layer: In general, a layer of air adjacent to a bounding surface. Specifically, the term most often refers to the planetary boundary layer, which is the layer within which the effects of friction are significant. From the earth, the layer is considered to be roughly the lowest one or two kilometers of the atmosphere.

bow echo: A radar echo that is linear but bent outward in a bow shape. Damaging straight-line winds often occur near the "crest" or center of a bow echo. Areas of circulation also can develop at either end of a bow echo, which sometimes can lead to tornado formation—especially in the left (usually northern) end, where the circulation exhibits cyclonic rotation.

box (or watch box): A severe thunderstorm or tornado watch.

breezy: Wind in the range of 15 mph to 25 mph with mild or warm temperatures.

brisk: Wind in the range of 15 to 25 mph when the temperature is cold.

broken clouds: Clouds that cover between five-eighths and seven-eighths of the sky.

Btu: British thermal unit, a measuring unit of heat.

Bulk Richardson number (or BRN): A non-dimensional number relating vertical stability and vertical shear (generally, stability divided by shear). High values indicate unstable and/or weakly sheared environments; low values indicate weak instability and/or strong vertical shear. Generally, values in the range of around 50 to 100 suggest environmental conditions favorable for supercell development.

buoyancy: That property of an object that enables it to float on the surface of a liquid or the rainy downdraft area of a thunderstorm or a complex of thunderstorms. A gust front or outflow boundary separates a bubble high from the surrounding air.

bust: An inaccurate forecast, usually a situation in which significant weather is expected but does not occur.

BWER (bounded weak echo region): Also known as a vault. Radar signature within a thunderstorm characterized by a local minimum in radar reflectivity at low levels that extends upward into, and is often surrounded by, higher reflectivities. This feature is associated with a strong updraft and is almost always found in the inflow region of a thunderstorm. It cannot be seen visually.

CAA: Cold air advection.

calm: The absence of apparent motion in the air.

calorie: The amount of heat required to raise the temperature of 1 gram of water 1°C.

cap (or capping inversion): A layer of relatively warm air aloft (usually several thousand feet above the ground) that suppresses or delays the development of thunderstorms. Air parcels rising into this layer become cooler than the surrounding air, which inhibits their ability to rise further. As such, the cap often prevents or delays thunderstorm development even in the presence of extreme instability.

CAPE: Convective available potential energy. A measure of the amount of energy available for convection. CAPE is directly related to the maximum potential vertical speed within an updraft; thus, higher values indicate greater potential for severe weather. Observed values in thunderstorm environments often may exceed 1,000 joules per kilogram (j/kg), and in extreme cases may exceed 5,000 j/kg. However, as with other indices or indicators, there are not threshold values above which severe weather becomes imminent.

carbon adsorption: Process whereby activated carbon, known as the sorbent, is used to remove certain wastes from water by preferentially holding them to the carbon surface.

carbon cycle: The atmosphere is a reservoir of gaseous carbon dioxide, but to be of use to life, this carbon dioxide must be converted into suitable organic compounds—"fixed"—as in the production of plant stems by the process of photosynthesis. The productivity of an area of vegetation is measured by the rate of carbon fixation. The carbon fixed by photosynthesis is eventually returned to the atmosphere as plants and animals die, and the dead organic matter is consumed by the decomposer organisms.

carbon dioxide: A colorless, odorless inert gas—a by-product of combustion.

carbon monoxide: A highly toxic and flammable gas that is a by-product of incomplete combustion. Very dangerous even in very low concentrations.

catabolism: In biology, the destructive part of metabolism where living tissue is changed into energy and waste products.

catalysis: The acceleration (or retardation) of chemical or biochemical reactions by a relatively small amount of a substance (the catalyst), which itself undergoes no permanent chemical change and which may be recovered when the reaction has finished.

catalyst: A substance or compound that speeds up the rate of chemical or biochemical reactions.

Cb: Cumulonimbus cloud.

ceiling: The height of the lowest layer of broken or overcast clouds.

cell: Convection in the form of a single updraft, downdraft, or updraft/downdraft couplet, typically seen as a vertical dome or tower as in a cumulus or towering cumulus cloud. A typical thunderstorm consists of several cells.

cellometer: A device used to evaluate the height of clouds or the vertical visibility into a surface-based obscuration.

Celsius: A temperature scale in which 0 is the freezing point of water and 100 is the boiling point.

CERCLA: Comprehensive Environmental Response, Compensation and Liability Act of 1980—aka, Superfund. CERCLA provides for cleanup and compensation and assigns liability for the release of hazardous substances into the air, land, or water.

chance: A 30%, 40%, or 50% chance of occurrence of measurable precipitation.

chemical bond: A chemical linkage that holds atoms together to form molecules.

chemical change: A transfer that results from making or breaking chemical bonds.

chemical equation: A shorthand method for expressing a reaction in terms of written chemical formulas.

chemical extraction: Process in which excavated contaminated soils are washed to remove contaminants of concern.

chemical formula: In the case of substances that consist of molecules, the chemical formula indicates the kinds of atoms present in each molecule and the actual number of them.

chemical oxygen demand (COD): A means of measuring the pollution strength of domestic and industrial wastes based upon the fact that all organic compounds, with few exceptions, can be oxidized by the action of strong oxidizing agents under acid conditions to carbon dioxide and water.

chemical precipitation: Process by which inorganic contaminants (heavy metals from groundwater) are removed by addition of carbonate, hydroxide, or sulfide chemicals.

chemical process audit/survey: A procedure used to gather information on the type, composition, and quantity of waste produced.

chemical reactions: When a substance undergoes a chemical change and is no longer the same substance; it becomes one or more new substances.

chemical weathering: A form of weathering brought about by a chemical change in the rocks affected; involves the breakdown of the minerals within a rock and usually produces a claylike residue.

chemosynthesis: A method of making protoplasm using energy from chemical reactions, in contrast to the use of light energy employed for the same purpose in photosynthesis.

chinook wind: A strong downslope wind that causes the air to warm rapidly as a result of compressive heating; called a foehn wind in Europe.

chlorofluorocarbons (CFCs): Synthetic chemicals that are odorless, nontoxic, nonflammable, and chemically inert. CFCs have been used as propellants in aerosol cans, as refrigerants in refrigeration and air conditioners, and in the manufacture of foam packaging. They are partly responsible for the destruction of the ozone layer.

circulation: The pattern of the movement of air. General circulation is the flow of air of large, semi-permanent weather systems, while secondary circulation is the flow of air of more temporary weather systems.

cirriform: High-altitude ice clouds with a very thin, wispy appearance.

cirrocumulus: Cirrus clouds with vertical development.

cirrostratus: Cirrus clouds with a flat, sheetlike appearance.

cirrus: High clouds, usually above 18,000 feet, composed of ice crystals and appearing in the form of white, delicate filaments or white or mostly white patches or narrow bands.

Clean Air Act: The name given to two acts passed by the U.S. Government. The Act of 1963 dealt with the control of smoke from industrial and domestic sources. It was extended by the Act of 1968, particularly to control gas cleaning and heights of stacks of installations in which fuels are burned to deal with smoke from industrial open bonfires. The 1990 Clean Air Act brought wide-ranging reforms for all kinds of pollution from large or small mobile or stationary sources, including routine and toxic emissions ranging from power plants to consumer products.

Clean Water Act (CWA): A keystone environmental law credited with significantly cutting the amount of municipal and industrial pollution fed into the nation's waterways. More formally known as the Federal Water Pollution Control Act Amendments, passed in 1972, it stems originally from a much-amended 1948 law aiding communities in building sewage treatment plants and has itself been much amended, most notably in 1977 and 1987.

clean zone: That point in a river or stream upstream before a single point of pollution discharge.

clear slot: A local region of clearing skies or reduced cloud cover, indicating an intrusion of drier air; often seen as a bright area with higher cloud bases on the west or southwest side of a wall cloud.

climate: The composite pattern of weather conditions that can be expected in a given region. Climate refers to yearly cycles of temperature, wind, rainfall, and so on, not to daily variations.

climate change: This strictly refers to all forms of climatic inconsistency. But it is often used in a more restricted sense to imply a significant change. Within the media, climate change has been used synonymously with global warming. Scientists, however, use the term in a wider sense to include past climate changes also.

climate normals: Averages of temperatures, precipitation, snowfall, and so on, made over standard 30-year periods. These normally span across three decades and are rederived every 10 years.

climatology: The scientific study of climate.

closed low: A low-pressure area with a distinct center of cyclonic circulation that can be completely encircled by one or more isobars or height contour lines. The term usually is used to distinguish a low-pressure area aloft from a low-pressure trough. Closed lows aloft typically are partially or completely detached from the main westerly current and thus move relatively slowly.

cloud: A visible cluster of tiny water and/or ice particles in the atmosphere.

cloud base: For a given cloud or cloud layer, it is the lowest level in the atmosphere where cloud particles are visible.

cloud condensation nuclei: Small particles in the air on which water vapor condenses and forms cloud droplets.

cloud streets: Rows of cumulus or cumulus-type clouds aligned parallel to the low-level flow. Cloud streets sometimes can be seen from the ground, but they are seen best on satellite photographs.

cloud tags: Ragged, detached cloud fragments; fractus or scud.

cloudburst: A sudden, intense rainfall that is normally of short duration.

cloudy: The state of the sky when seven-tenths or more of the sky is covered by clouds.

coastal flood warning: Issued when there is widespread coastal flooding expected within 12 hours, more than just typical overwash.

coastal flooding: The inundation of land areas along the coast caused by sea water above normal tidal actions. This is often caused by prolonged strong onshore flow of wind and/or high astronomical tides.

coastal forecast: A forecast of wind, wave, and weather conditions between the coastline and 25 miles offshore.

coastal waters: Includes the area from a line approximating the mean high water along the mainland or island as far out as 25 miles including the bays, harbors, and sounds.

cold advection (CAA): Transport of cold air into a region by horizontal winds.

cold-air damming: Cold air damming occurs when a cold dome of high pressure settles over northeastern New England. The clockwise circulation around the high pressure center brings northeasterly winds to the mid-Atlantic region. The northeasterly winds bank cold air against the eastern slopes of the Appalachian Mountains. Warmer air from the west or southwest is lifted above the cold air as it moves instead of warming the surface.

cold-air funnel: A funnel cloud or (rarely) a small, relatively weak tornado that can develop from a small shower or thunderstorm when the air aloft is unusually cold (hence the name). They are much less violent than other types of tornadoes.

cold front: The leading portion of a cold atmospheric air mass moving against and eventually replacing a warm air mass.

cold low: A low-pressure system with a cold air mass from near the surface to all vertical levels.

cold pool: A region of relatively cold air, represented on a weather map analysis as a relative minimum in temperature surrounded by closed isotherms. Cold pools aloft represent regions of relatively low stability, while surface-based cold pools are regions of relatively stable air.

collar cloud: Frequently used as a synonym for wall cloud, although it is actually a generally circular ring of cloud surrounding the upper portion of a wall cloud.

combined seas: The combined height of swell and wind waves.

combustion: In chemical terms, the rapid combination of a substance with oxygen, accompanied by the evolution of heat and usually light. In air pollution control, combustion or incineration is a beneficial pollution control process in which the objective is to convert certain contaminants to innocuous substances such as carbon dioxide and water.

comma cloud: A synoptic-scale cloud pattern with a characteristic comma-like shape, often seen on satellite photographs associated with large and intense low-pressure systems.

commercial chemical products: An EPA category listing of hazardous wastes (also called *P* or *U* listed wastes because their code numbers begin with these letters); includes specific commercial chemical products or manufacturing chemical intermediates.

commercial sources of MSW: Solid waste generated in restaurants, hotels, stores, motels, service stations, repair shops, markets, office buildings, and print shops.

Community Right-to-Know Act: A part of SARA Title III under CERCLA. Stipulates that a community located near a facility storing, producing, or using hazardous materials has a right to know about the potential consequences of a catastrophic chemical spill or release of chemicals from the site.

composite sample: A sample formed by mixing discrete samples taken at periodic points in time or a continuous proportion of the flow. The number of discrete samples that make up the composite depends upon the variability of pollutant concentration and flow.

composting: A beneficial reuse biological process whereby waste (e.g., yard trimmings or wastewater biosolids) is transformed into a harmless humus-like substance used as a soil amendment.

compound: A substance composed of two or more elements, chemically combined in a definite proportion.

concentrated solution: Solute in concentration present in large quantities.

condensation: Air pollution control technology used to remove gaseous pollutants from waste streams; a process in which the volatile gases are removed from the contaminant stream and changed into a liquid.

condensation nuclei: Small particles in the air around which water vapor condenses.

condenser: Air pollution control device used in a condensation method to condense vapors to a liquid phase either by increasing the system pressure without a change in temperature or by decreasing the system temperature to its saturation temperature without a pressure change.

conduction: Heat flow of heat energy through a material without the movement of any part of the material itself.

confluence: A pattern of wind flow in which air flows inward toward an axis oriented parallel to the general direction of flow. It is the opposite of difluence. Confluence is not the same as convergence. Winds often accelerate as they enter a confluent zone, resulting in speed divergence that offsets the (apparent) converging effect of the confluent flow.

congestus (or cumulus congestus): A large cumulus cloud with great vertical development, usually with a cauliflower-like appearance but lacking the characteristic anvil-shaped top of a cumulonimbus cloud.

continental air mass: A dry air mass originating over a larger land area.

contrail: A cloud-like stream formed in cold, clear air behind the engines of an airplane.

convection: Method of heat transfer whereby the heated molecules circulate through the medium (gas or liquid).

convective outlook: A forecast containing the area(s) of expected thunderstorm occurrence and expected severity over the contiguous United States, issued several times daily by the SPC.

convective temperature: The approximate temperature that the air near the ground must warm to in order for surface-based convection to develop, based on analysis of a sounding.

convergence: An atmospheric condition that exists when the winds cause a horizontal net inflow of air into a specified region. Divergence is the opposite, where winds cause a horizontal net outflow of air from a specified region.

cooling degree day: A form of degree day used to estimate the required energy for cooling. One cooling degree day occurs for each degree the daily mean temperature is above 65°F.

Coordinated Universal Time (UTC): The time in the 0° meridian time zone.

Coriolis force: An apparent force caused by the rotation of the earth. In the northern hemisphere winds are deflected to the right, and in the southern hemisphere to the left. In synoptic-scale weather systems (hurricanes and large mid-latitude storms), the Coriolis force causes the air to rotate around a low-pressure center in a cyclonic direction. The air flowing around a hurricane spins counterclockwise in the northern hemisphere.

corona: A disk of light surrounding the sun or moon; this is a result of the diffraction of light by small water droplets.

corrosive: A substance that attacks and eats away other materials by strong chemical action.

covalent bond: A chemical bond produced when two atoms share one or more pairs of electrons.

CRS: Console Replacement System. This consists of a computer system and computer voice that is used to automate NOAA Weather Radio.

cumulonimbus cloud (Cb): A vertically developed cloud, often capped by an anvil-shaped cloud. Also called a thunderstorm cloud, it is frequently accompanied by heavy showers, lightning, thunder, and sometimes hail or gusty winds.

cumulus cloud: A cloud in the shape of individual detached domes, with a flat base and a bulging upper portion resembling cauliflower.

cumulus congestus: A large cumulus cloud with great vertical development, usually with a cauliflower-like appearance but lacking the characteristic anvil-shaped top of a cumulonimbus cloud.

cut-off low: An upper-level low-pressure system that is no longer in the normal west-to-east upper airflow. Usually a cut-off low will lie to the south of the established upper airflow.

cyclogenesis: Development or intensification of a low-pressure center.

cyclone: (1) In air pollution control, a cyclone collector is used to remove particles from a gas stream by centrifugal force; (2) an area of low pressure around which winds blow counterclockwise in the northern hemisphere; (3) also the term used for a hurricane in the Indian Ocean and in the western Pacific Ocean.

cyclonic circulation (or cyclonic rotation): Circulation (or rotation) that is in the same sense as the earth's rotation—that is, counterclockwise (in the northern hemisphere) as would be seen from above.

cytochrome: A class of iron-containing proteins important in cell metabolism.

cytoplasm: The jelly-like matter within a cell.

dart leader: In lightning, the leader that, after the first stroke, initiates each succeeding stroke of a composite flash of lightning.

debris cloud: A rotating "cloud" of dust or debris, near or on the ground, often appearing beneath a condensation funnel and surrounding the base of a tornado.

decomposers: Organisms such as bacteria, mushrooms, and fungi that obtain nutrients by breaking down complex matter in the wastes and dead bodies of other organisms into simpler chemicals, most of which are returned to the soil and water for reuse by producers.

decomposition: Process whereby a chemical compound is reduced to its component substances. In biology, the destruction of dead organisms either by chemical reduction or by the action of decomposers.

decouple: The tendency for the surface wind to become much lighter than wind above it at night when the surface temperature cools.

degree day: A measure of the departure of the daily mean temperature from the normal daily temperature; heating and cooling degree days are the departure of the daily mean temperature from 65°F.

dendrite: Hexagonal ice crystals with complex and often fernlike branches.

dense fog: A fog in which the visibility is less than one-quarter mile.

dense fog advisory: Issued when fog is expected to reduce visibility to a quarter mile or less over a widespread area for at least three hours.

density: The ratio of the weight of a mass to the unit of volume.

density altitude: The pressure altitude corrected for temperature deviations from the standard atmosphere. It is used by pilots when setting aircraft performance.

density of air: The mass of air divided by its volume. The air's density depends on its temperature, its pressure, and how much water vapor is in the air.

depletion: In evaluating ambient air quality, pertains to the fact that pollutants emitted into the atmosphere do not remain there forever.

depression: A region of low atmospheric pressure that is usually accompanied by low clouds and precipitation.

depth hoar: Large (one to several millimeters in diameter), cohesionless, coarse, faceted snow crystals that result from the presence of strong temperature gradients within the snowpack.

derecho: A widespread and usually fast-moving windstorm associated with convection. Derechoes include any family of downburst clusters produced by an extratropical MCS; they can produce damaging straight-line winds over areas hundreds of miles long and more than 100 miles across.

detoxification: Biological conversion of a toxic substance to one less toxic.

dew: Moisture from water vapor in the air that has condensed on objects near the ground, whose temperatures have fallen below the dew point temperature.

dew point: The temperature to which the air must be cooled for water vapor to condense and form fog or clouds.

diamond dust: A fall of non-branched (snow crystals are branched) ice crystals in the form of needles, columns, or plates.

differential motion: Cloud motion that appears to differ relative to other nearby cloud elements (e.g., clouds moving from left to right relative to other clouds in the foreground or background). Cloud rotation is one example of differential motion, but not all differential motion indicates rotation. For example, horizontal wind shear along a gust front may result in differential cloud motion without the presence of rotation.

diffusion: (1) Mixing of substances, usually gases and liquids, from molecular motion; (2) the spreading out of a substance to fill a space.

difluence (or diffluence): A pattern of wind flow in which air moves outward (in a fan-out pattern) away from a central axis that is oriented parallel to the general direction of the flow. It is the opposite of confluence.

dilute solutions: A solution weakened by the addition of water, oil, or other liquid or solid.

dirty ridge: Most of the time, upper-level ridges bring fairly clear weather as the storms are steered around the ridge. Sometimes, however, strong storms undercut the ridge and create precipitation. Ridges that experience this undercutting by storms are known as dirty ridges because of the unusual precipitation.

disinfection: Effective killing by chemical or physical processes all organisms capable of causing infectious disease.

dispersion: The dilution and reduction of concentration of pollutants in either air or water. Air pollution dispersion mechanisms are a function of the prevailing meteorological conditions.

disturbance: A disruption of the atmosphere that usually refers to a low pressure area, cool air, and inclement weather.

diurnal: Daily; related to actions that are completed in the course of a calendar day and that typically recur every calendar day (e.g., diurnal temperature rises during the day and falls at night).

divergence: The expansion or spreading out of a vector field; usually said of horizontal winds. It is the opposite of convergence.

doldrums: The regions on either side of the equator where air pressure is low and winds are light.

Doppler radar: A type of weather radar that determines whether atmospheric motion is toward or away from the radar. It determines the intensity of rainfall and uses the Doppler effect to measure the velocity of droplets in the atmosphere.

downburst: A strong downdraft resulting in an outward burst of damaging wind on or near the ground. Downburst winds can produce damage similar to a strong tornado.

downdraft: A column of generally cool air that rapidly sinks to the ground, usually accompanied by precipitation as in a shower or thunderstorm.

downslope wind: Air that descends an elevated plain and consequently warms and dries. Occurs when prevailing wind direction is from the same direction as the elevated terrain and often produces fair weather conditions.

downstream: In the same direction as a stream or other flow, or toward the direction in which the flow is moving.

drifting snow: Uneven distribution of snowfall caused by strong surface winds. Drifting snow does not reduce visibility.

drizzle: Small, slowly falling water droplets, with diameters between 0.2 and 0.5 millimeters.

drought: Abnormally dry weather in a region over an extended period sufficient to cause a serious hydrological (water cycle) imbalance in the affected area. This can cause such problems as crop damage and water-supply shortage.

dry adiabat: A line of constant potential temperature on a thermodynamic chart.

dry adiabatic lapse rate: When a dry parcel of air is lifted in the atmosphere, it undergoes adiabatic expansion and cooling that results in a lapse rate (cooling) of $-1°C/100m$ or $-10°C/km$.

dry line: A boundary separating moist and dry air masses, an important factor in severe weather frequency in the Great Plains.

dry-line bulge: A bulge in the dry line, representing the area where dry air is advancing most strongly at lower levels.

dry punch: A surge of drier air; normally a synoptic-scale or mesoscale process. A dry punch at the surface results in a dry-line bulge.

dry slot: A zone of dry (and relatively cloud-free) air that wraps east-northeastward into the southern and eastern parts of a synoptic scale or mesoscale low-pressure system. A dry slot generally is seen best on satellite photographs.

dust devil: A small, rapidly rotating wind that is made visible by the dust, dirt, or debris it picks up; also called a whirlwind. Dust devils usually develop during hot, sunny days over dry and dusty or sandy areas.

dust plume: A non-rotating "cloud" of dust raised by straight-line winds; often seen in a microburst or behind a gust front.

dust storm: An area where high surface winds have picked up loose dust, reducing visibility to less than one-half mile.

dust whirl: A rotating column of air rendered visible by dust.

dynamics: Generally, any forces that produce motion or effect change. In operation meteorology, dynamics usually refer specifically to those forces that produce vertical motion in the atmosphere.

easterly wave: A wavelike disturbance in the tropical easterly winds that usually moves from east to west. Such waves can grow into tropical depressions.

ECMG: European Center for Meteorology Forecast Model.

ecological toxicology: The branch of toxicology that addresses the effect of toxic substances, on not only the human population but also the environment in general, including air, soil, surface water, and groundwater.

ecology: The study of the interrelationship of an organism or a group of organisms with their environment.

ecosystem: A self-regulating natural community of plants and animals interacting with one another and with their nonliving environment.

ecotoxicology: See *ecological toxicology.*

eddy: A small volume of air that behaves differently from the predominant flow of the layer in which it exists, seemingly having a life of its own. An example of such would be a tornado, which has its own distinct rotation but is different from the large-scale flow of air surrounding the thunderstorm in which the tornado is born.

El Niño: A major warming of the equatorial waters in the eastern Pacific Ocean. El Niño events usually occur every 3 to 7 years and are related to shifts in global weather patterns. (Spanish for the "Christ Child"; named this because it often begins around Christmas.)

electron: A component of an atom; travels in a distant orbit around a nucleus.

electrostatic precipitation: Process using a precipitator to remove dust or other particles from air and other gases by electrostatic means. An electric discharge is passed through the gas, giving the impurities a negative electric charge. Positively charged plates are then used to attract the charged particles and remove them from the fast flow.

elements: The simplest substance that cannot be separated into more simple parts by ordinary means. There are more than 100 known elements.

endergonic: A reaction in which energy is absorbed.

energy: A system capable of producing a physical change of state.

enhanced greenhouse effect: The natural greenhouse effect has been enhanced by man's emissions of greenhouse gases. Increased concentrations of carbon dioxide, methane, and nitrous oxide trap more infrared radiation, so heating up of the atmosphere occurs.

enhanced wording: An option used by the SPC in tornado and severe thunderstorm watches when the potential for strong/violent tornadoes, or unusually widespread damaging straight-line winds, is high.

entrance region: The region upstream from a wind speed maximum in a jet stream (jet max), in which air is approaching (entering) the region of maximum winds and therefore is accelerating. This acceleration results in a vertical circulation that creates divergence in the upper-level winds in the right half of the entrance region (as would be viewed looking along the direction of flow). This divergence results in upward motion of air in the right rear quadrant (or right entrance region) of the jet max. Severe weather potential sometimes increases in this area as a result.

ENSO: El Niño / Southern Oscillation.

entropy: A measure of the disorder of a system.

environment: All the surroundings of an organism, including other living things, climate and soil, and so on; in other words, the conditions for development or growth.

environmental degradation: All the limiting factors that act together to regulate the maximum allowable size or carrying capacity of a population.

environmental factors: Factors that influence volatilization of hydrocarbon compounds from soils. Environmental factors include temperature, wind, evaporation, and precipitation.

environmental science: The study of the human impact on the physical and biological environment of an organism. In its broadest sense, it also encompasses the social and cultural aspects of the environment.

environmental toxicology: The branch of toxicology that addresses the effect of toxic substances, not only on the human population but also on the environment in general, including air, soil, surface water, and groundwater.

enzymes: Proteinaceous substances that catalyze microbiological reactions such as decay or fermentation. They are not used up in the process but speed it up greatly. They can promote a wide range of reactions, but a particular enzyme can usually promote a reaction only on a specific substrate.

equilibrium level (or EL): On a sounding, the level above the level of free convection (LFC) at which the temperature of a rising air parcel again equals the temperature of the environment.

ETA: "Eta" (from Greek) is a model generated every 12 hours by NCEP.

evaporation: The process of a liquid changing into a vapor or gas.

evapotranspiration: Combination of evaporation and transpiration of liquid water in plant tissue and in the soil to water vapor in the atmosphere.

excessive heat warning: Issued within 12 hours of the onset of the following conditions: heat index of at least 105°F for more than 3 hours per day for 2 consecutive days or heat index of more than 115°F for any period of time.

excessive heat watch: Issued for the potential of the following conditions within 12 to 36 hours: heat index of at least 105°F for more than 3 hours per day for 2 consecutive days or heat index of more than 115°F for any period of time.

exergonic: Releasing energy.

exit region: The region downstream from a wind speed maximum in a jet steam (jet max), in which air is moving away from the region of maximum winds and therefore is decelerating. This deceleration results in divergence in the upper-level winds in the left half of the exit region (as would be viewed looking along the direction of flow). This divergence results in upward motion of air in the left front quadrant (or left exit region) of the jet max. Severe weather potential sometimes increases in this area as a result.

extended outlook: A basic forecast of general weather conditions three to five days in the future.

extratropical cyclone: A storm that forms outside the tropics, sometimes as a tropical storm or hurricane changes.

eye: The low-pressure center of a tropical cyclone. Winds are normally calm and sometimes the sky clears.

eye wall: The ring of thunderstorms that surrounds a storm's eye. The heaviest rain, strongest winds, and worst turbulence are normally in the eye wall.

Fahrenheit: The standard scale used to measure temperature in the United States, in which the freezing point of water is 32° and the boiling point is 212°.

fair: Describes weather in which there is less than four-tenths of opaque cloud cover, no precipitation, and no extreme visibility, wind, or temperature conditions.

fall wind: A strong, cold, downslope wind.

feeder bands: Lines or bands of low-level clouds that move (feed) into the updraft region of a thunderstorm, usually from the east through south (i.e., parallel to the inflow). This term also is used in tropical meteorology to describe spiral-shaped bands of convection surrounding, and moving toward, the center of a tropical cyclone.

fetch: The area in which ocean waves are generated by the wind. Also refers to the length of the fetch area, measured in the direction of the wind.

few: A cloud layer that covers between one-eighth and two-eighths of the sky.

first law of thermodynamics: In any chemical or physical change, movement of matter from one place to another, or change in temperature, energy is neither created nor destroyed but merely converted from one form to another.

flanking line: A line of cumulus connected to and extending outward from the most active portion of a parent cumulonimbus, usually found on the southwest side of the storm. The cloud line has roughly a stair-step appearance with the taller clouds adjacent to the parent cumulonimbus. It is most frequently associated with strong or severe thunderstorms.

flash flood: A flood that occurs within a few hours (usually less than six) of heavy or excessive rainfall, dam or levee failure, or water released from an ice jam.

flash flood warning: Issued to inform the public, emergency management, and other cooperating agencies that flash flooding is in progress, imminent, or highly likely.

flash flood watch: Issued to indicate current or developing hydrologic conditions that are favorable for flash flooding in and close to the watch area, but the occurrence is neither certain nor imminent.

flood: A condition that occurs when water overflows the natural or artificial confines of a stream or river; the water also may accumulate by drainage over low-lying areas.

flood crest: The highest stage or flow occurring in a flood.

flood stage: The stage at which water overflowing the banks of a river, stream, or body of water begins to cause damage.

flood warning: Issued when there is expected inundation of a normally dry area near a stream or other water course or unusually severe ponding of water.

flurries: Light snow falling for short durations. No accumulation or just a light dusting is all that is expected.

foehn: A warm, dry wind on the lee side of a mountain range. The heating and drying are due to adiabatic compression as the wind descends downslope.

fog: Water that has condensed close to ground level, producing a cloud of very small droplets that reduces visibility to less than one kilometer (3,300 feet).

fogbow: A rainbow that has a white band that appears in fog and is fringed with red on the outside and blue on the inside.

forecast: A forecast provides a description of the most significant weather conditions expected during the current and following days. The exact content depends upon the intended user, such as the public or marine forecast audiences.

formula weight: The sum of the atomic weight of all atoms that comprise one formula unit.

fractus: Ragged, detached cloud fragments.

freeze: Occurs when the surface air temperature is expected to be 32°F or below over a widespread area for a significant period of time.

freeze warning: Issued during the growing season when surface temperatures are expected to drop below freezing over a large area for an extended period of time, regardless if frost develops or not.

freezing: The change in a substance from a liquid to a solid state.

freezing drizzle: Drizzle that falls in liquid form and then freezes upon impact with the ground or an item with a temperature of 32°F or less, possibly producing a thin coating of ice. Even in small amounts, freezing drizzle may cause traveling problems.

freezing fog: A suspension of numerous minute ice crystals in the air, or water droplets at temperatures below 0°C, based at the earth's surface, which reduces horizontal visibility; also called ice fog.

freezing level: The altitude in the atmosphere where the temperature drops to 32°F.

freezing nuclei: Particles suspended in the air around which ice crystals form.

freezing rain: Rain that freezes on objects such as trees, cars, and roads, forming a coating or glaze of ice. Temperatures at higher levels are warm enough for rain to form, but surface temperatures are below 32°F, causing the rain to freeze on impact.

freshet: The annual spring rise of streams in cold climates as a result of snow melt; freshet also refers to a flood caused by rain or melting snow.

friable: Readily crumbled in the hand.

frog storm: The first bad weather in spring after a warm period.

front: In meteorology, the boundary between two air masses of different temperatures or humidity. The basic frontal types are cold fronts, warm fronts, and occluded fronts.

frost: The formation of thin ice crystals on the ground or other surfaces. Frost develops when the temperature of the exposed surface falls below 32°F and water vapor is deposited as a solid.

frost advisory: Issued during the growing season when widespread frost formation is expected over an extensive area. Surface temperatures are usually in the mid-30s Fahrenheit.

frost dew: When liquid dew changes into tiny beads of ice. The change occurs after dew formation when the temperature falls below freezing.

frost point: When the temperature to which air must be cooled to in order to be saturated is below freezing.

Fujita scale: System developed by Dr. Theodore Fujita to classify tornadoes based on wind damage. Scale is from F0 for weakest to F5 for strongest tornadoes.

Fujiwhara effect: The Fujiwhara effect describes the rotation of two storms around each other.

fumigation: Results when emissions from a smokestack that is under an inversion layer head downward, leading to greatly elevated downwind ground-level concentrations of contamination.

funnel cloud: A rotating, cone-shaped column of air extending downward from the base of a thunderstorm but not touching the ground. When it reaches the ground it is called a tornado.

gale: Sustained wind speeds from 34 to 47 knots (39 to 54 mph).

gale warning: A marine weather warning for gale-force winds from a nontropical system.

gas: In the widest sense, applied to all aeriform bodies, the most minute particles of which exhibit the tendency to fly apart from each other in all directions. Normally these gases are found in that state at ordinary temperature and pressure. They can be liquefied or solidified only by artificial means, either through high pressure or extremely low temperatures.

gas laws: The physical laws concerning the behavior of gases. They include Boyle's law and Charles's law, which are concerned with the relationships between the pressure, temperature, and volume of an ideal (hypothetical) gas.

geosphere: Consists of the inorganic, or nonliving, portions of earth, which are home to all the globe's organic, or living, matter.

geostationary satellite: A satellite positioned over the equator that rotates at the same rate as the earth, remaining over the same spot.

glaciation: The transformation of cloud particles from water droplet to ice crystals. Thus, a cumulonimbus cloud is said to have a "glaciated" upper portion.

glaze: A layer or coating of ice that is generally smooth and clear and forms on exposed objects by the freezing of liquid raindrops.

global warming: The long-term rise in the average temperature of the earth.

GOES: Geostationary Operational Environmental Satellite.

GOES-8: One of the Geostationary Operational Environmental Satellites. They are owned and run by the National Oceanic and Atmospheric Administration (NOAA), while NASA designs and launches them.

grab sample: An individual discrete sample collected over a period of time not exceeding 15 minutes.

gradient: The time rate or spatial rate of change of an atmospheric property.

gram: The basic unit of weight in the metric system; equal to 1/1000 of a kilogram; approximately 28.5 grams equal one ounce.

graupel: Small pellets of ice created when supercooled water droplets coat, or rime, a snowflake. The pellets are cloudy or white, not clear like sleet, and often are mistaken for hail.

gravity: The force of attraction that arises between objects by virtue of their masses. On earth, gravity is the force of attraction between any object in the earth's gravitational field and the earth itself.

gravity wave: A wave disturbance in which buoyancy acts as the restoring force on parcels displaced from hydrostatic equilibrium. Waves on the ocean are examples of gravity waves.

greenhouse effect: The trapping of heat in the atmosphere. Incoming short-wavelength solar radiation penetrates the atmosphere, but the longer-wavelength outgoing radiation is absorbed by water vapor, carbon dioxide, ozone, and several other gases in the atmosphere and is reradiated to earth, causing an increase in atmospheric temperature.

greenhouse gases: The gases present in the earth's atmosphere that cause the greenhouse effect.

ground fog: Shallow fog (less than 20 feet deep) produced over the land by the cooling of the lower atmosphere as it comes in contact with the ground; also known as radiation fog.

growing degree day: A form of degree day to estimate the approximate dates when a crop will be ready to harvest; one growing degree day occurs when the daily mean temperature is 1° above the minimum temperature required for the growth of that specific crop.

growing season: The period of time between the last killing frost of spring and the first killing frost of autumn.

gust: A brief sudden increase in wind speed. Generally the duration is less than 20 seconds and the fluctuation greater than 10 mph.

gust front: The leading edge of the downdraft from a thunderstorm. A gust front may precede the thunderstorm by several minutes and have winds that can easily exceed 80 mph.

gustnado (or gustinadao): Gust front tornado. A small tornado, usually weak and short-lived, that occurs along the gust front of a thunderstorm. Often it is visible only as a debris cloud or dust while near the ground.

hail: Precipitation in the form of balls or irregular lumps of ice produced by liquid precipitation, freezing, and being coated by layers of ice as it is lifted and cooled in strong updrafts of thunderstorms.

halo: A ring or arc that encircles the sun or moon. Halos are caused by the refraction of light through the ice crystals in cirrus clouds.

hard freeze: A freeze where vegetation is killed and the ground surface is frozen solid.

harmattan: A hot, dry, and dusty northeasterly or easterly wind that occurs in West Africa north of the equator and is caused by the outflow of air from subtropical high-pressure areas.

haze: Fine dust or salt particles in the air that reduce visibility.

heat: A condition of matter caused by the rapid movement of its molecules. Energy has to be applied to the material in sufficient amounts to create the motion and may be applied by mechanical or chemical means.

heat advisory: Issued within 12 hours of the onset of the following conditions: heat index of at least 105°F but less than 115°F for less than 3 hours per day. Nighttime lows that remain above 80°F for 2 consecutive days.

heat balance: The constant trade-off that takes place when solar energy reaches the earth's surface and is absorbed, then must return to space to maintain earth's normal heat balance.

heat index: An index that combines air temperature and humidity to give an apparent temperature (how hot it feels).

heat islands: Large metropolitan areas where heat generated has an influence on the ambient temperature (adds heat) in and near the area.

heat lightning: Lightning that can be seen but is too far away for the thunder to be heard.

heating degree day: A form of degree day used to estimate the required energy for heating. One heating degree day occurs for each degree the daily mean temperature is below 65°F.

heavy snow: Depending on the region of the United States, this generally means that 4 or more inches of snow has accumulated in 12 hours, or 6 or more inches of snow in 24 hours.

heavy snow warning: Older terminology replaced by winter storm warning for heavy snow. Issued when 7 or more inches of snow or sleet is expected in the next 24 hours. A warning is used for winter weather conditions posing a threat to life and property.

heavy surf: The result of large waves breaking on or near the shore resulting from swells or produced by a distant storm.

helicity: A property of a moving fluid that represents the potential for helical flow (i.e., flow that follows the pattern of a corkscrew) to evolve. Helicity is proportional to the strength of the flow, the amount of vertical wind shear, and the amount of turning in the flow (i.e., vorticity).

Henry's law: Governs the behavior of gases in contact with water.

high: An area of high pressure, usually accompanied by anticyclonic and outward wind flow; also known as an anticyclone.

high risk (of severe thunderstorms): Severe weather is expected to affect more than 10% of the area.

high wind warning: Issued when sustained winds from 40 to 73 mph are expected for at least 1 hour or when any wind gusts are expected to reach 58 mph or more.

high wind watch: Issued when conditions are favorable for the development of high winds over all of or part of the forecast area but the occurrence is still uncertain. The criteria of a high wind watch are listed under the high wind warning and should include the area affected, the reason for the watch, and the potential impact of the winds.

hodograph: A plot representing the vertical distribution of horizontal winds, using polar coordinates. A hodograph is obtained by plotting the end points of the wind vectors at various altitudes and connecting these points in order of increasing height.

hook echo: A radar pattern sometimes observed in the southwest quadrant of a tornadic thunderstorm. Appearing like a fishhook turned in toward the east, the hook echo is precipitation aloft around the periphery of a rotation column of air 2 to 10 miles in diameter.

horse latitudes: Subtropical regions where anticyclones produce settled weather.

hot spot: Typically large areas of pavement, these "hot spots" are heated much quicker by the sun than surrounding grasses and forests. As a result, air rises upwards from the relatively hot surface of the pavement, reaches its condensation level, condenses, and forms a cloud above the "hot spot."

humidity: The amount of water vapor in a given volume of the atmosphere (absolute humidity), or the ratio of the amount of water vapor in the atmosphere to the saturation value at the same temperature (relative humidity).

hurricane: A severe tropical cyclone with sustained winds over 74 mph (64 knots); normally applied to such storms in the Atlantic Basin and the Pacific Ocean east of the international date line.

hurricane warning: Warning issued when sustained winds of 74 mph (64 knots) or more are expected within 24 hours. This implies a dangerous storm surge.

hydrocarbon: A chemical containing only carbon and hydrogen atoms. Crude oil is a mixture largely of hydrocarbons.

hydrological cycle: The means by which water is circulated in the biosphere. Cooling in the atmosphere and precipitation over both land and oceans counterbalances evapotranspiration from the land mass plus evaporation from the oceans.

hydrology: The study of the waters of the earth with relation to the effects of precipitation and evaporation upon the water in streams, rivers, and lakes and its effect on land surfaces.

hydrosphere: The portion of the earth's surface covered by the oceans, seas, and lakes.

hygrometer: An instrument used to measure humidity.

ice age: Periods in the history of the earth characterized by a growth of the ice caps toward the equator and a general lowering of global surface temperatures, especially in temperate mid-latitudes. The most recent ice age ended about 10,000 years ago. Ice advances in this period are known to have altered the whole pattern of global atmospheric circulation.

ice crystals: A barely visible crystalline form of ice that has the shape of needles, columns, or plates. Ice crystals are so small that they seem to be suspended in air. Ice crystals occur at very low temperatures (around 0°F and colder) in a stable atmosphere.

ice fog: A suspension of numerous minute ice crystals in the air, or water droplets at temperatures below 0°C, based at the earth's surface, which reduces horizontal visibility. Usually occurs at –20°F and below.

ice jam: An accumulation of broken river ice caught in a narrow channel that frequently produces local floods during a spring break-up.

ice pellets: Precipitation of transparent or translucent pellets of ice, which are round or irregular, are rarely conical, and have a diameter of 0.2 inch (5 mm) or less. There are two main types: hard grains of ice consisting of frozen raindrops and pellets of snow encased in a thin layer of ice.

ice storm: Liquid rain falling and freezing on contact with cold objects, creating ice buildups of 14 inches or more that can cause severe damage.

ice storm warning: Older terminology replaced by winter storm warning for severe icing. Issued when half an inch or more of accretion of freezing rain is expected. This may lead to dangerous walking or driving conditions and the pulling down of power lines and trees. A warning is used for winter weather conditions posing a threat to life and property.

ideal gas law: A hypothetical gas that obeys the gas laws exactly in regard to temperature, pressure, and volume relationships.

impaction: In air pollution control technology, a particle collection process whereby the center mass of a particle diverging from a fluid strikes a stationary object and is collected by the stationary object.

in situ technologies: Remedial technologies performed in place at the site.

indefinite ceiling: The ceiling classification applied when the reported ceiling value represents the vertical visibility upward into surface-based obscuration.

Indian summer: An unseasonably warm period near the middle of autumn, usually following a substantial period of cool weather.

inflow bands (or feeder bands): Bands of low clouds, arranged parallel to the low-level winds and moving into or toward a thunderstorm.

inflow jets: Local jets of air near the ground flowing inward toward the base of a tornado.

inflow notch: A radar signature characterized by an indentation in the reflectivity pattern on the inflow side of the storm. The indentation often is V-shaped, but this term should not be confused with V-notch. Supercell thunderstorms often exhibit inflow notches, usually in the right quadrant of a classic supercell, but sometimes in the eastern part of an HP storm or in the rear part of a storm (rear inflow notch).

inflow stinger: A beaver-tail cloud with a stinger-like shape.

infrared radiation: Invisible electromagnetic radiation of a wavelength between about 0.75 micrometers and 1 mm—between the limit of the red end of the visible spectrum and the shortest microwaves.

inorganic substance: A substance that is mineral in origin that does not contain carbon compounds, except as carbonates, carbides, and so on.

insolation: The amount of direct solar radiation incident per unit of horizontal area at a given level.

instability: A state of the atmosphere in which convection takes place spontaneously, leading to cloud formation and precipitation.

Intertropical Convergence Zone (ITCZ): The region where the northeasterly and southeasterly trade winds converge, forming an often continuous band of clouds or thunderstorms near the equator.

inversion: An increase in temperature with height; the reverse of the normal cooling with height in the atmosphere. Temperature inversions trap atmospheric pollutants in the lower troposphere, resulting in higher concentrations of pollutants at ground levels than would usually be experienced.

ionic bonds: A chemical bond in which electrons have been transferred from atoms of low ionization potential to atoms of high electron affinity.

ionosphere: Also known as the thermosphere; a layer in the atmosphere above the mesosphere extending from about 80 km above the earth's surface. It can be considered a distinct layer due to a rise in air temperature with increasing height. Atmospheric densities here are very low.

iridescence: Brilliant patches of green or pink sometimes seen near the edges of high- or medium-level clouds.

isentropic lift: Lifting of air that is traveling along an upward-sloping isentropic surface. Situations invoking isentropic lift often are characterized by widespread stratiform clouds and precipitation.

isentropic surface: A two-dimensional surface containing points of equal potential temperature.

isobar: A line drawn on maps and weather charts linking all places with the same atmospheric pressure (usually measured in millibars).

isodrosotherm: A line of equal dew point temperature.

isohyet: A line of equal precipitation amounts.

isopleth: General term for a line of equal value of some quality. Isobars, isotherms, and so on are all examples of isopleths.

isotach: A line of equal wind speed.

isotherm: A line of equal temperature on a weather map.

January thaw: A period of mild weather popularly supposed to recur each year in later January.

jet streak: A local wind speed maximum within a jet stream.

jet stream: A narrow band of very fast wind found at altitudes of 6 to 10 miles in the upper troposphere or lower stratosphere.

katabatic: Wind blowing down an incline, such as down a hillside; downslope wind.

katafront: A front (usually a cold front) at which the warm air descends the frontal surface.

Kelvin: Temperature scale used by scientists that begins at absolute zero and increases by the same degree intervals as the Celsius scale; that is, 0°C is the same as 273K and 100°C is 373K.

killing frost: Frost severe enough to end the growing season. This usually occurs at temperatures below 28°F.

kilopascal: The internationally recognized unit for measuring atmospheric pressure. It is equal to 10 millibars.

knot: A measure of speed. It is one nautical mile per hour (1.15 mph). A nautical mile is one minute of 1° of latitude.

knuckles: Lumpy protrusions on the edges, and sometimes the underside, of a thunderstorm anvil. They usually appear on the upwind side of a back-sheared anvil, indicating rapid expansion of the anvil due to the presence of a very strong updraft. They are not mammatus clouds.

lake effect: The effect of a lake (usually a large one) in modifying the weather near the shore and downwind. It often refers to the enhanced rain or snow that falls downwind from the lake. This effect can also result in enhanced snowfall along the east coast of New England in winter.

laminar: Smooth, non-turbulent; often used to describe cloud formations that appear to be shaped by a smooth flow of air traveling in parallel layers or sheets.

La Niña: A cooling of the equatorial waters in the Pacific Ocean.

land breeze: A wind that blows from the land toward a body of water; also known as an offshore breeze. It occurs when the land is cooler than the water.

landspout: A tornado that does not arise from organized storm-scale rotation and, therefore, is not associated with a wall cloud (visually) or a mesocyclone (on radar). Landspouts typically are observed beneath cumulonimbus or towering cumulus clouds (often as not more than a dust whirl) and essentially are the land-based equivalents of weather spouts.

lapse rate: The rate of change of air temperature with increasing height in the atmosphere.

latent heat: The heat energy that must be absorbed when a substance changes from solid to liquid and liquid to gas and that is released when a gas condenses and a liquid solidifies.

latent heat of fusion: The amount of heat required to change 1 gram of a substance from the solid to the liquid phase at the same temperature.

latent heat of vaporization: The amount of heat required to change 1 gram of a substance from the liquid to the gas phase at the same temperature.

Law of conservation of mass: In any ordinary physical or chemical change, matter is neither created nor destroyed but merely changed from one form to another.

layer: An array of clouds and/or obscurations whose bases are at approximately the same level.

leeward: Situated away from the wind; downwind—opposite of windward.

left front quadrant (or left exit region): The area downstream from and to the left of an upper-level jet max (as would be viewed looking along the direction of flow). Upward motion and severe thunderstorm potential sometimes are increased in this area relative to the wind speed maximum.

left mover: A thunderstorm that moves to the left relative to the steering winds and to other nearby thunderstorms; often the northern part of a splitting storm.

lenticular clouds: A cloud that generally has the form of a smooth lens. They usually appear in formation as the result of orographic origin. Viewed from the ground, the clouds appear stationary as the air rushes through them.

lifted index (or LI): A common measure of atmospheric instability. Its value is obtained by computing the temperature that air near the ground would have if it were lifted to some higher level (around 18,000 feet, usually) and comparing that temperature to the actual temperature at that level. Negative values indicate instability—the more negative, the more unstable the air is; and if thunderstorms develop, they are more likely to be stronger.

lifting: The forcing of air in a vertical direction by the upslope in terrain or by the movement of a denser air mass.

lifting condensation level: The level in the atmosphere where a lifted air parcel reaches its saturation point, and as a result the water vapor within condenses into water droplets.

lightning: Any form of visible electrical discharges produced by thunderstorms.

likely: In probability of precipitation statements, the equivalent of a 60% to 70% chance.

limiting factor: Factors such as temperature, light, water, or chemicals that limit the existence, growth, abundance, or distribution of an organism.

liquid: A state of matter between a solid and a gas.

liter: A metric unit of volume equal to one cubic decimeter (1.76 pints).

lithosphere: The earth's crust; the layers of soil and rock that comprise the earth's crust.

loaded ground (sounding): A sounding characterized by extreme instability but containing a cap, such that explosive thunderstorm development can be expected if the cap can be weakened or the air below it heated sufficiently to overcome it.

long-wave trough: A trough in the prevailing westerly flow aloft that is characterized by large length and (usually) long duration. Generally, there are no more than about five long-wave troughs around the northern hemisphere at any given time. Their position and intensity govern general weather patterns (e.g., hot/cold, wet/dry) over periods of days, weeks, or months.

low: An area of low pressure, usually accompanied by cyclonic and inward wind flow; also known as a cyclone.

low-level jet: A region of relatively strong winds in the lower part of the atmosphere.

macroburst: Large downburst with a 2.5-mile or greater outflow diameter and damaging winds lasting 5 to 20 minutes.

mamma clouds: Also called mammatus, these clouds appear as hanging, rounded protuberances of pouches on the undersurface of a cloud. With thunderstorms, mammatus are seen on the underside of the anvil. These clouds do not produce tornadoes, funnels, hail, or any other type of severe weather, although they often accompany severe thunderstorms.

maritime air mass: An air mass that forms over water. It is usually humid and may be cold or warm.

mass: The quantity of matter and a measurement of the amount of inertia that a body possesses.

mass balance equations: Used to track pollutants from one place to another.

materials balance: The law of conservation of mass/matter that says that everything has to go somewhere but is neither created nor destroyed in the process.

maximum temperature: The highest temperature during a specified time period.

mean sea level (MSL): The average height of the sea surface, based upon hourly observation of the tide height on the open coast or in adjacent waters that have free access to the sea.

mean temperature: The average of a series of temperatures taken over a period of time, such as a day or month.

medium range: In forecasting, (generally) three to seven days in advance.

melting point: The temperature at which a substance changes from solid to liquid.

mercury barometer: An instrument that measures barometric pressure by measuring the level of mercury in a column.

meridional flow: A type of atmospheric circulation pattern in which the north and south component of motion is unusually pronounced; opposite of zonal flow.

mesocyclone: A storm-scale region of rotation, typically around 2 to 6 miles in diameter and often found in the right rear flank of a supercell (or often on the eastern, or front, flank of an HP storm). The circulation of a mesocyclone covers an area much larger than the tornado that may develop within it.

mesohigh: A mesoscale high-pressure area, usually associated with MCSs or their remnants.

mesolow (or sub-synoptic low): A mesoscale low-pressure center. Severe weather potential often increases in the area near and just ahead of a mesolow.

mesonet: A regional network of observing stations (usually surface stations) designed to diagnose mesoscale weather features and their associated processes.

mesoscale: Size scale referring to weather systems smaller than synoptic-scale systems but larger than single storm clouds. Horizontal dimensions generally range from around 50 miles to several hundred miles. Squall lines are an example of mesoscale weather systems.

mesoscale convective complex (MCC): A large mesoscale convective system, generally round or oval-shaped, that normally reaches peak intensity at night. The formal definition includes specific minimum criteria for size, duration, and eccentricity (i.e., "roundness"), based on the cloud shield as seen on infrared satellite photographs.

mesoscale convective system (MCS): A complex of thunderstorms that becomes organized on a scale larger than the individual thunderstorms and normally persists for several hours or more. MCSs may be round or linear in shape and include systems such as tropical cyclones, squall lines, and MCCs (among others). MCS often is used to describe a cluster of thunderstorms that do not satisfy the size, shape, or duration criteria of an MCC.

mesosphere: An atmospheric layer that extends from the top of the stratosphere to about 56 miles above the earth.

META: The mesoscale ETA model; a mathematical model of the atmosphere run on a computer that makes forecasts out to 30 hours.

metamorphism: Changes in the structure and texture of snow grains that result from variations in temperature, migration of liquid water and water vapor, and pressure within the snow cover.

METAR: A weather observation near ground level. It may include date and time, wind, visibility, wealth and obstructions to vision, sky condition, temperature and dew point, sea level pressure, precipitation amount, and other data used for aircraft operations.

meteorologist: A person who studies meteorology. Some examples include research meteorologist, climatologist, operational meteorologist, and TVB meteorologist.

meteorology: The scientific observation and study of the atmosphere so that weather can be accurately forecast.

meter: The standard of length in the metric system, equal to 39.37 inches or 3.28 feet.

methane (CH_4): The simplest hydrocarbon of the paraffin series. Colorless, odorless, and lighter than air, it burns with a bluish flame and explodes when mixed with air or oxygen. Methane is a greenhouse gas.

microbiology: The study of organisms that can be seen only under the microscope.

microburst: A strong localized downdraft from a thunderstorm with peak gusts lasting two to five minutes.

microclimate: A local climate that differs from the main climate around it.

middens: Primitive dunghills or refuse heaps.

mid-latitudes: The areas in the northern and southern hemispheres between the tropics and the Arctic and Antarctic circles.

midnight dumping: The illegal dumping of solid or hazardous wastes into the environment.

millibar: A metric unit of atmospheric pressure; 1 mb = 100 Pa (pascal). Normal surface pressure is approximately 1,013 millibars.

minimum temperature: The lowest temperature during a specified time period.

mist: Consists of microscopic water droplets suspended in the air that produce a thin grayish veil over the landscape. It reduces visibility to a lesser extent than fog.

mixing: Air movements (usually vertical) that make the properties of the air with a parcel homogeneous. It may result in a lapse rate approaching the moist or dry adiabatic rate.

mixture: In chemistry, a substance containing two or more compounds that still retain their separate physical and chemical properties.

mobile sources: Non-stationary sources of gaseous pollutants, including locomotives, automobiles, ships, and airplanes.

modeling: Refers to the use of mathematical representations of contaminant dispersion and transformation to estimate ambient pollutant concentrations.

moderate risk (of severe thunderstorms): Severe thunderstorms are expected to affect between 5% and 10% of the area.

moisture advection: Transport of moisture by horizontal winds.

moisture convergence: A measure of the degree to which moist air is converging into a given area, taking into account the effect of converging winds and moisture advection. Areas of persistent moisture convergence are factored regions for thunderstorm development, if other factors (e.g., instability) are favorable

molar concentration (molarity): In chemistry, solution that contains one mole of a substance per liter of solvent.

mole: SI unit (symbol mol) of the amount of a substance; the amount of a substance that contains as many elementary entities as there are atoms in 12 grams of the isotope carbon-12.

molecular weight: The weight of one molecule of a substance relative to 12C, expressed in grams.

molecule: The fundamental particle that characterizes a compound. It consists of a group of atoms held together by chemical bonds.

monitoring: Process whereby a contaminant is tracked.

monsoon: A persistent seasonal wind, often responsible for a seasonal precipitation regime. It is most commonly used to describe meteorological changes in southern and eastern Asia.

Montreal Protocol: Required signatory countries to reduce their consumption of CFCs by 20% by 1993 and by 50% by 1998.

morning glory: An elongated cloud band, visually similar to a roll cloud, usually appearing in the morning hours, when the atmosphere is relatively stable. Morning glories result from perturbations related to gravitational waves in a stable boundary layer.

MOS: Model Output Statistics.

mountain breeze: System of winds that blow downhill during the night.

MRF: Medium range forecast model generated every 12 hours by NCEP.

MSL: Mean sea level.

MSLP: Mean sea level pressure.

muggy: Colloquially descriptive of warm and especially humid weather.

multicell cluster thunderstorm: A thunderstorm consisting of two or more cells, of which most or all are often visible at a given time as distinct domes or towers in various stages of development.

multiple-vortex tornado: A tornado in which two or more condensation funnels or debris clouds are present at the same time, often rotating about a common center or about each other. Multiple-vortex tornadoes can be especially damaging.

mushroom: A thunderstorm with a well-defined anvil rollover and thus having a visual appearance resembling a mushroom.

National Ambient Air Quality Standards (NAAQS): Established by the EPA at two levels: primary and secondary. Primary standards must be set at levels that will protect public health and include an "adequate margin of safety," regardless of whether the standards are economically or technologically achievable. Primary standards must protect even the most sensitive individuals, including the elderly and those with respiratory ailments. Secondary air quality standards are meant to be even more stringent than primary standards. Secondary standards are established to protect public welfare (e.g., structures, crops, animals, or fabrics).

NCDC: National Climatic Data Center. Located in Asheville, North Carolina, this agency archives climatic and forecast data from the National Weather Service.

NCEP: National Centers for Environmental Prediction; central computer and communications facility of the National Weather Service; located in Washington, DC.

negative-tilt trough: An upper-level system that is tilted to the west with increasing latitude (i.e., with an axis from southeast to northwest). A negative-tilt trough often is a sign of a developing or intensifying system.

neutrally stable atmosphere: An intermediate class between stable and unstable conditions. Will cause a smokestack plume to cone in appearance as the edges of the plume spread out in a V-shape.

neutron: Elementary particles that have approximately the same mass as protons but have no charge. They are one constituent of the atomic nucleus.

NEXRAD: NEXt Generation RADar; a National Weather Service network of about 140 Doppler radars operating nationwide.

NGM: Nested grid model generated every 12 hours by NCEP.

NHC: National Hurricane Center; the office of the national Weather Service in Miami that is responsible for tracking and forecasting tropical cyclones.

nitrogen cycle: The natural circulation of nitrogen through the environment.

nitrogen dioxide (NO^2): A reddish-brown, highly toxic gas with a pungent odor; one of the seven known nitrogen oxides that participate in photochemical smog and primarily affect the respiratory system.

nitrogen fixation: Nature accomplishes nitrogen fixation by means of nitrogen-fixing bacteria.

nitrogen oxide (NO): A colorless gas used as an anesthetic.

NOAA: National Oceanic and Atmospheric Administration. A branch of the U.S. Department of Commerce, NOAA is the parent organization of the National Weather Service.

NOAA Weather Radio (NWR): Continuous, 24-hour-a-day VHF broadcasts of weather observations and forecasts directly from National Weather Service offices. A special tone allows certain receivers to alarm when watches or warnings are issued.

NOAA Weather Wire (NWWS): A computer dissemination network that sends National Weather Service products to the media and public.

nocturnal: Related to nighttime or occurring at night.

nonpoint source: Source of pollution in which wastes are not released at one specific, identifiable point but from a number of points that are spread out and difficult to identify and control.

nonpoint source pollution: Pollution that cannot be traced to a specific source but rather comes from multiple generalized sources.

nonvolatile: A substance that does not evaporate at normal temperatures when exposed to the air.

Nor'easter: A low-pressure disturbance forming along the South Atlantic coast and moving northeast along the Middle Atlantic and New England coasts to the Atlantic provinces of Canada. It usually causes strong northeast winds with rain or snow. Also called a Northeaster or coastal storm.

normal: The long-term average value of a meteorological element for a certain area. For example, "temperatures are normal for this time of year," usually averaged over 30 years.

normal lapse rate: The rate of temperature change with height is called the lapse rate. On average, temperature decreases –65°C/100m or –6.5°C/km, the normal lapse rate.

northern lights: Also known as the aurora borealis. The luminous radiation emission from the upper atmosphere over middle and high latitudes, centered around the earth's magnetic poles. These silent fireworks are often seen on clear winter nights in a variety of shapes and colors.

nowcast: A short-term weather forecast, generally out to six hours or less.

NSSL: National Severe Storms Laboratory.

nucleus: A particle of any nature upon which molecules or water or ice accumulate.

numerical forecasting: Forecasting the weather through digital computations carried out by supercomputers.

nutrient cycles: See *biogeochemical cycles.*

nutrients: Elements or compounds needed for the survival, growth, and reproduction of a plant or animal.

NWP: Numerical weather prediction.

NWS: National Weather Service.

obscuration: Any phenomenon in the atmosphere, other than precipitation, that reduces the horizontal visibility in the atmosphere.

occluded front: A complex frontal system that occurs when a cold front overtakes a warm front; also known as an occlusion.

offshore breeze: A wind that blows from the land toward a body of water; also known as a land breeze.

offshore forecast: A marine weather forecast for the waters between 60 and 250 miles off the coast.

omega: A term used to describe vertical motion in the atmosphere. The "omega equation" used in numerical weather models is composed of two terms, the "differential vorticity advection" term and the "thickness advection" term. Put more simply, omega is determined by the amount of spin (or large-scale rotation) and warm (or cold) advection present in the atmosphere. On a weather forecast chart, high values of omega (or a strong omega field) relate to upward vertical motion in the atmosphere. If this upward vertical motion is strong enough and in a sufficiently moist air mass, precipitation results.

onshore breeze: A wind that blows from a body of water toward the land; also known as a sea breeze.

organic chemistry: The branch of chemistry concerned with compounds of carbon.

organic matter: Includes both natural and synthetic molecules containing carbon and usually hydrogen. All living matter is made up of organic molecules.

organic substance: Any substance containing carbon.

orographic: Related to, or caused by, physical geography (such as mountains or sloping terrain).

orographic lift: The lifting of air as it passes over terrain features such as hills or mountains. This can create orographic clouds and/or precipitation.

orphan anvil: An anvil from a dissipated thunderstorm, below which no other clouds remain.

outflow: Air that flows outward from a thunderstorm.

outflow boundary: A storm-scale or mesoscale boundary separating thunderstorm-cooled air (outflow) from the surrounding air; similar in effect to a cold front, with passage marked by a wind shift and usually a drop in temperature.

outflow winds: Winds that blow down fjords and inlets from the land to the sea.

overcast: Sky condition when greater than nine-tenths of the sky is covered by clouds.

overrunning: A condition that exists when a relatively warm air mass moves up and over a colder and denser air mass on the surface. The result is usually low clouds, fog, and steady, light precipitation.

overshooting top (or penetrating top): A dome-like protrusion above a thunderstorm anvil, representing a very strong updraft and hence a higher potential for severe weather with that storm.

oxidation: The process by which electrons are lost.

oxidation-reduction: The (redox) process where electrons are lost and gained.

oxidize: To combine with oxygen.

oxygen: An element that readily unites with materials.

ozone: The compound O_3. Found naturally in the atmosphere in the ozonosphere, a constituent of photochemical smog.

ozone holes: Holes created in the ozone layer because of chemicals, especially CFCs.

particulate matter: Normally refers to dust and fumes; travels easily through air.

partly cloudy: Sky condition when between three-tenths and seven-tenths of the sky is covered; used more frequently at night.

partly sunny: Similar to partly cloudy; used to emphasize daytime sunshine.

pascal (Pa): A unit of pressure equal to one newton per square meter.

patches: Used with fog to denote random occurrence over relatively small areas.

pathogen: Any disease-producing organism.

pendant echo: Radar signature generally similar to a hook echo, except that the hook shape is not as well defined.

periodic law: The properties of elements are periodic functions of the atomic number.

periodic table: A list of all elements arranged in order of increasing atomic numbers and grouped by similar physical and chemical characteristics into "periods"; based on the chemical law that physical or chemical properties of the elements are periodic functions of their atomic weights.

permafrost: A soil layer below the surface of tundra regions that remains frozen permanently.

perpetual resource: A resource such as solar energy that comes from an essentially inexhaustible source and thus will always be available on a human time scale regardless of whether or how it is used.

persistent substance: A chemical product with a tendency to persist in the environment for quite some time—plastics, for example.

pesticide: Any chemical designed to kill weeds, insects, fungi, rodents, and other organisms that humans consider to be undesirable.

pH: A numerical designation of relative acidity and alkalinity; a pH of 7.0 indicates precise neutrality, high values indicate increasing alkalinity, and lower values indicate increasing acidity.

photochemical reaction: A reaction induced by the presence of light.

photochemical smog: A complex mixture of air pollutants produced in the atmosphere by the reaction of hydrocarbons and nitrogen oxides under the influence of sunlight.

photosynthesis: A complex process that occurs in the cells of green plants whereby radiant energy from the sun is used to combine carbon dioxide (CO_2) and water (H_2O) to produce oxygen (O_2) and simple sugar or food molecules, such as glucose.

plume: (1) The column of non-combustible products emitted from a fire or smokestack; (2) a vapor cloud formation having shape and buoyancy; (3) a contaminant formation dispersing through the subsurface.

point source: Discernible conduits, including pipes, ditches, channels, sewers, tunnels, or vessels, from which pollutants are discharged.

point source pollution: Pollution that can be traced to an identifiable source.

polar air: A mass of very cold, very dry air that forms in polar regions.

polar front: The semi-permanent, semi-continuous front that encircles the northern hemisphere, separating air masses of tropical and polar origin.

polar stratospheric clouds (PSCs): High-altitude clouds that form in the stratosphere above Antarctica during the southern hemisphere winter. Their presence seems to initiate the ozone loss experienced during the ensuing southern hemisphere spring.

polar vortex: A circumpolar wind circulation that isolates the Antarctic continent during the cold southern hemisphere winter, heightening ozone depletion.

pollute: To impair the quality of some portion of the environment by the addition of harmful impurities.

polycrystal: A snowflake composed of many individual ice crystals.

POP: Probability of precipitation. Probability forecasts are subjective estimates of the chances of encountering measurable precipitation at some time during the forecast period.

popcorn convection: Clouds, showers, and thundershowers that form on a scattered basis with little or no apparent organization, usually during the afternoon in response to diurnal heating.

positive area: The area on a sounding representing the layer in which a lifted parcel would be warmer than the environment; thus, the area between the environmental temperature profile and the path of the lifted parcel.

positive-tilt trough: An upper-level system that is tilted to the east with increasing latitude (i.e., from southwest to northeast). A positive-tilt trough often is a sign of a weakening weather system and generally is less likely to result in severe weather than a negative-tilt trough if all other factors are equal.

potential temperature: The temperature a parcel of dry air would have if brought adiabatically (i.e., without transfer of heat or mass) to a standard pressure level of 1,000 millibars.

precipitation: Liquid or solid water that falls from the atmosphere and reaches the ground.

precipitation shaft: A visible column of rain and/or hail falling from a cloud base. When viewed against a light background, heavy precipitation appears very dark gray, sometimes with a turquoise tinge. This turquoise tinge has been commonly attributed to hail, but its actual cause is unknown.

pressure: Force per unit area.

pressure change: The net difference between pressure readings at the beginning and ending of a specified interval of time.

pressure falling rapidly: A decrease in station pressure at a rate of 0.06 inch of mercury or more per hour that totals 0.02 inch or more.

pressure gradient force: A variation of pressure with position.

pressure rising rapidly: An increase in station pressure at a rate of 0.06 inch of mercury or more per hour that totals 0.02 inch or more.

pressure tendency: The character and amount of atmospheric pressure change during a specified period of time, usually the three-hour period preceding an observation.

pressure unsteady: A pressure that fluctuates by 0.03 inch of mercury or more from the mean pressure during the period of measurement.

prevailing westerlies: Winds in the middle latitudes (approximately 30° to 60°) that generally blow from west to east.

prevailing wind: The direction from which the wind blows most frequently in any location.

primary pollutants: Pollutants emitted directly into the atmosphere, where they exert an adverse influence on human health or the environment. The six primary pollutants are carbon dioxide, carbon monoxide, sulfur oxides, nitrogen oxides, hydrocarbons, and particulates. All but carbon dioxide are regulated in the United States.

primary standards: The Clean Air Act (NAAQS) air quality standard covering criteria pollutants.

profiler: An instrument designed to measure horizontal winds directly above its location and thus to measure the vertical wind profile. Profilers operate on the same principles as Doppler radar.

proton: A component of a nucleus, 2,000 times more massive than an electron; differs from a neutron by its positive (+1) electrical charge. The atomic number of an atom is equal to the number of protons in its nucleus.

psychrometer: An instrument used for measuring the water vapor content of the atmosphere. It consists of two thermometers, one of which is an ordinary glass thermometer, while the other has its bulb covered with a jacket of clean muslin that is saturated with distilled water prior to use.

pulse storm: A thunderstorm within which a brief period (pulse) of strong updraft occurs, during and immediately after which the storm produces a short episode of severe weather. These storms generally are not tornado producers but often produce large hail and/or damaging winds.

PVA: Positive vorticity advection. Advection of higher values of vorticity into an area, which often is associated with upward motion (lifting) of the air. PVA typically is found in advance of disturbances (i.e., short waves) and is a property that often enhances the potential precipitation.

QFP: Quantitative precipitation forecast.

quality of snow: The amount of ice in a snow sample expressed as a percentage of the weight of the sample.

radar: An instrument used to detect precipitation by measuring the strength of the electromagnetic signal reflected back (RADA = radio detection and ranging).

radiation: The emitting of energy from an atom in the form of particles of electromagnetic waves; energy waves that travel with the speed of light and upon arrival at a surface are either absorbed, reflected, or transmitted.

radiation fog: Fog produced over the land by the cooling of the lower atmosphere as it comes in contact with the ground; also known as ground fog.

radiational cooling: Cooling process of the earth's surface and adjacent air that occurs when infrared (heat) energy radiates from the surface of the earth upward through the atmosphere into space. Air near the surface transfers its thermal energy to the nearby ground through conduction so that radiative cooling lowers the temperature of both the surface and the lowest part of the atmosphere.

radiative inversions: In temperature inversions, a nocturnal phenomenon caused by cooling of the earth's surface. Inversions prompt the formation of fog and simultaneously trap gases and particulates, creating a concentration of pollutants.

radioactive material: Any material that spontaneously emits ionizing radiation.

radiosonde: An instrument attached to a weather balloon that transmits pressure, humidity, temperature, and winds as it ascends to the upper atmosphere.

rain: Liquid water droplets that fall from the atmosphere, having diameters greater than drizzle (0.5 mm).

rainbow: Optical phenomenon when light is refracted and reflected by moisture in the air into concentric arcs of color. Raindrops act like prisms, breaking the light into the colors of a rainbow, with red on the outer and blue on the inner edge.

rain foot: A horizontal bulging near the surface in a precipitation shaft, forming a foot-shaped prominence. It is a visual indication of a wet microburst.

rain-free base: A horizontal, dark cumulonimbus base that has no visible precipitation beneath it. This structure usually marks the location of the thunderstorm updraft. Tornadoes most commonly develop (1) from wall clouds that are attached to the rain-free base or (2) from the rain-free base itself. This is particularly true when the rain-free base is observed to the south or southwest of the precipitation shaft.

rain gauge: An instrument used to measure rainfall amounts.

rain shadow: The region on the lee side of a mountain or mountain range where the precipitation is noticeably less than on the windward side.

Rankine temperature scale: A temperature scale with the degree of the Fahrenheit temperature scale and the zero point of the Kelvin temperature scale.

rawinsonde: A balloon that is tracked by radar to measure wind speeds and wind directions in the atmosphere.

reactive: The tendency of a material to react chemically with other substances.

reduction: Removal of oxygen from a compound; lowering of the oxidation number resulting from a gain of electrons.

reflectivity: Radar term referring to the ability of a radar target to return energy; used to estimate precipitation intensity and rainfall rates.

refraction: The bending of light as it passes through areas of different density, such as from air through ice crystals.

relative humidity: The percentage of moisture in a given volume of air at a given temperature in relation to the amount of moisture the same volume of air would contain at the saturation point.

renewable resources: Resources that can be depleted in the short run if used or contaminated too rapidly but that normally are replaced through natural processes.

representative sample: A sample of a universe or whole, such as a waste pile, lagoon, or groundwater, that can be expected to exhibit the average properties of the whole.

resource: Something that serves a need and is useful and available at a particular cost.

retrogression (or retrograde motion): Movement of a weather system in a direction opposite to that of the basic flow in which it is embedded, usually referring to a closed low or a longwave trough that moves westward.

return flow: South winds on the back (west) side of an eastward-moving surface high-pressure system. Return flow over the central and eastern United States typically results in a return of moist air from the Gulf of Mexico (or the Atlantic Ocean).

RFC: River Forecast Center. The Northeast River Forecast Center is located in Taunton, Massachusetts.

ridge: An elongated area of high pressure in the atmosphere; opposite of a trough.

right entrance region (or right rear quadrant): The area upstream from and to the right of an upper-level jet max (as would be viewed looking along the direction of flow). Upward motion and severe thunderstorm potential sometimes are increased in this area relative to the wind speed maximum.

right mover: A thunderstorm that moves appreciably to the right relative to the main steering winds and to other nearby thunderstorms. Right movers typically are associated with a high potential for severe weather. (Supercells often are right movers.)

rime: Tiny balls of ice that form when tiny drops of water (usually not precipitation) freeze on contact with the surface.

river flood warning: Issued when main stem rivers (such as the Merrimack, Charles, Connecticut, etc.) are expected to reach a level above flood stage.

roll cloud: A relatively rare, low-level, horizontal, tube-shaped accessory cloud completely detached from the cumulonimbus base. When present, it is located along the gust front and most frequently observed on the leading edge of a line of thunderstorms. The roll cloud will appear to be slowly "rolling" about its horizontal axis. Roll clouds are not and do not produce tornadoes.

rope (or rope funnel): A narrow, often contorted condensation funnel usually associated with the decaying stage of a tornado.

rope cloud: In satellite meteorology, a narrow, rope-like band of clouds sometimes seen on satellite images along a front or other boundary.

rope stage: The dissipating stage of a tornado, characterized by thinning and shrinking of the condensation funnel into a rope (or a rope funnel). Damage still is possible during this stage.

Rossby waves: Long waves that form in air or water that flows almost parallel to the equator, which results from the effect of the earth's rotation.

rotor cloud: A turbulent cloud formation found in the lee of some large mountain barriers. The air in the cloud rotates around an axis parallel to the mountain range.

RUC: Rapid Update Cycle; a numerical model run at NCEP that focuses on short-term (up to 12 hours) forecasts and small-scale (mesoscale) weather features. Forecasts are prepared every 3 hours for the contiguous United States.

runway visual range (RVR): An instrumentally derived value, based on standard calibrations, that represents the horizontal distance a pilot may see down the runway from the approach end.

Saffir-Simpson Hurricane Damage Potential Scale: A scale that measures hurricane intensity, developed by Herbert Saffir and Robert Simpson.

sandstorm: Particles of sand carried aloft by a strong wind. The sand particles are mostly confined to the lowest 10 feet and rarely rise more than 50 feet above the ground

Santa Anna winds: Relatively warm, dry winds that blow into Southern California coastal areas from an anticyclone located over the high deserts of California or Nevada. The warmth and dryness are due to compressional heating.

satellite photo: A photograph of the earth taken by weather satellites that shows areas of cloud.

saturation: A condition of the atmosphere in which a certain volume of air holds the maximum water vapor it can hold at a specific temperature.

saturation vapor pressure (water): The maximum amount of water vapor necessary to keep moist air in equilibrium with a surface of pure water. This is the maximum water vapor the air can hold for any given combination of temperature and pressure.

scattered: A cloud layer that covers between three-eighths and one-half of the sky.

science: The observation, identification, description, experimental investigation, and theoretical explanation of natural phenomena.

scientific method: A systematic form of inquiry that involves observation, speculation, and reasoning.

scud clouds: Small, ragged, low cloud fragments that are unattached to a larger cloud base and often seen with and behind cold fronts and thunderstorm gust fronts. Such clouds generally are associated with cool moist air, such as thunderstorm outflow.

sea breeze: A wind that blows from a sea or ocean toward a land mass; also known as an onshore breeze. It occurs when the land is warmer than the water.

sea-level pressure: The pressure value obtained by the theoretical reduction or increase of barometric pressure to sea level.

sea-level rise: The natural rise of sea level that occurs in cyclical patterns throughout history; may be the result of man's impact on global warming.

second law of thermodynamics: Natural law that dictates that in any conversion of heat energy to useful work, some of the initial energy input is always degraded to a lower-quality, more dispersed, less useful form of energy, usually low-temperature heat that flows into the environment; you can't break even in terms of energy quality.

secondary cold front: A front that follows a primary cold front and ushers in even colder air.

secondary standards: Refer to NAAQS requirements to protect public welfare.

sensible heat: The excess radiative energy that has passed from the earth's surface to the atmosphere through advection, conduction, and convection processes.

severe thunderstorm: A strong thunderstorm with wind gusts in excess of 58 mph (50 knots) and/or hail with a diameter of 0.75 inch or more.

severe thunderstorm warning: Issued when thunderstorms are expected to have wind gusts to 58 mph or above or hail 0.75 inch or more in diameter.

severe thunderstorm watch: Issued when conditions are favorable for the development of severe thunderstorms in and close to a defined area.

shallow fog: Fog in which the visibility at 6 feet above ground level is five-eighths of a mile or more.

shear (wind shear): Variation in wind speed and/or direction over a short distance. Shear usually refers to vertical wind shear—that is, the change in wind with height; but the term also is used in Doppler radar to describe changes in radial velocity over short horizontal distances.

shelf cloud: A low-level horizontal accessory cloud that appears to be wedge-shaped as it approaches. It is usually attached to the thunderstorm base and forms along the gust front. The leading edge of the shelf is often smooth and at times layered or terraced. It is most often seen along the leading edge of an approaching line of thunderstorms, accompanied by gusty straight winds as it passes overhead and followed by precipitation. The underside is concave upward, turbulent, boiling, or wind-torn. Tornadoes rarely occur with the shelf cloud.

short-fuse warning: A warning issued by the NWS for a local weather hazard of relatively short duration. Short-fuse warnings include tornado warnings, severe thunderstorm warnings, and flash flood warnings. Tornado and severe thunderstorm warnings typically are issued for periods of an hour or less, flash flood warnings typically for three hours or less.

short wave (short-wave trough): A disturbance in the mid or upper part of the atmosphere that induces upward motion ahead of it. If other conditions are favorable, the upward motion can contribute to thunderstorm development ahead of a short wave.

short-wave radiation: The radiation received from the sun and emitted in spectral wavelengths less than 4 microns. It is also called "solar radiation."

shower: Precipitation that is intermittent, in time, space, or intensity.

sinks: Areas, whether natural or artificial, where the products or effluents from production and consumption in one place are physically exported to another for storage or dispersal.

sky condition: The state of the sky in terms of such parameters as sky cover, layers and associated heights, ceiling, and cloud types.

sky cover: The amount of the sky that is covered by clouds or obscurations in contact with the surface.

sleet: Sleet drops that freeze into ice pellets before reaching the ground. Sleet usually bounces when hitting a surface and does not stick to objects. Forms when snow enters a warm layer of air above the surface and melts, and then enters a deep layer of sub-freezing air near the surface and refreezes.

slight chance: In probability of precipitation statements, usually equivalent to a 20% chance.

slight risk (of severe thunderstorms): Severe thunderstorms are expected to affect between 2% and 5% of the area. A slight risk generally implies that severe weather events are expected to be isolated.

sling psychrometer: A psychrometer in which the wet and dry bulb thermometers are mounted upon a frame connected to a handle. The psychrometer may be whirled by hand in order to provide the necessary ventilation.

small craft advisory: A marine advisory for winds 25 to 33 knots (29 to 38 mph) or seas of 5 feet or more that may cause hazardous conditions for operators of small vessels.

smog: Term used to describe visible air pollution; a dense, discolored haze containing large quantities of soot, ash, and gaseous pollutants such as sulfur dioxide and carbon dioxide.

smoke: A suspension in the air of small particles produced by combustion. A transition to haze may occur when smoke particles have traveled great distances (25 to 100 statute miles or more) and when the larger particles have settled out and the remaining particles have become widely scattered through the atmosphere.

snow: Frozen precipitation composed of ice particles in complex hexagonal patterns. Snow forms in cold clouds by the direct transfer of water vapor to ice.

snow advisory: Older terminology replaced by winter weather advisory. An advisory issued when 4, 5, or 6 inches of snow or sleet is expected in 24 hours. It is expected to create hazardous or restricted travel conditions, but not as severe as expected with a winter storm.

snow depth: The vertical height of frozen precipitation on the ground. For this purpose, frozen precipitation includes ice pellets, glaze, hail, any combination of these, and sheet ice formed directly or indirectly from precipitation.

snow flurries: Light snow showers, usually of an intermittent nature and short duration with no measurable accumulation.

snow grains: Precipitation of very small, white, opaque grains of ice.

snow pellets: Precipitation of white, opaque grains of ice. The grains are round or sometimes conical. Diameters range from about 0.08 to 0.2 inch (2 to 5 mm).

snow shower: Snow falling at varying intensities for brief periods of time. Some accumulation is possible.

snow squalls: Intense, but of limited duration, periods of moderate to heavy snowfall, accompanied by strong, gusty surface winds and possible lightning.

snowburst: Very intense shower of snow, often of short duration, that greatly restricts visibility and produces periods of rapid snow accumulation.

snowfall: The depth of new snow that has accumulated since the previous day or since the previous observation.

snowflake: White ice crystals that have combined in a complex branched hexagonal form.

solar energy: The energy produced by the sun.

solid: Matter that has a definite volume and a definite shape.

solubility: The ability of a substance to mix with water.

solute: The dissolved substance in a solution.

solvent: The substance in excess in a solution.

sorption: Process of adsorption or absorption of a substance on or in another substance.

sounder: A special kind of radiometer that measures changes in atmospheric temperature with height, as well as the content of various chemical species in the atmosphere at various levels. The High Resolution Infrared Radiation Sound (HIRS), found on NOAA polar-orbiting satellites, is a passive instrument.

sounding: A plot of the vertical profile of temperature and dew point (and other winds) above a fixed location. Soundings are used extensively in weather forecasting, for example, to determine instability, locate temperature inversions, and so on.

southern oscillation: A periodic reversal of the pressure pattern across the tropical Pacific Ocean during El Niño events.

special marine warning: Issued for brief or sudden occurrence of sustained wind or frequent gusts of 34 knots or more. This is usually associated with severe thunderstorms or waterspouts.

specific heat: The amount of heat energy in calories necessary to raise the temperature of 1 gram of the substance by 1°C.

SPC: Storm Prediction Center. Located in Norman, OK, this office is responsible for monitoring and forecasting severe convective weather in the continental United States. This includes the issuance of tornado and severe thunderstorm watches.

speed shear: The component of wind shear that is due to a change in wind speed with height (e.g., southwesterly winds of 20 mph at 10,000 feet increasing to 50 mph at 20,000 feet). Speed shear is an important factor in severe weather development, especially in the middle and upper levels of the atmosphere.

spin-up: A small-scale vortex initiation, as may be seen when a gustnado, landspout, or suction vortex forms.

spray: An ensemble of water droplets torn by the wind from an extensive body of water, generally from the crests of waves, and carried up into the air in such quantities that it reduces the horizontal visibility.

squall: A strong wind characterized by a sudden onset in which the wind speed increases at least 16 knots and is sustained at 22 knots or more for at least one minute.

squall line: Any non-frontal line or narrow band of active thunderstorms. The term is usually used to describe solid or broken lines of strong or severe thunderstorms.

St. Elmo's Fire: A luminous, and often audible, electric discharge that is intermediate in nature between a spark discharge and point discharge (with its diffuse, quiescent, and non-luminous character). It occurs from objects, especially pointed ones, when the electric field strength near their surfaces attains a value near 100,000 volts per meter. Aircraft flying through active electrical storms often develop corona discharge streamers from antennas and propellers, and even from the entire fuselage and wing structure. It is seen also during stormy weather, emanating from the yards and masts of ships at sea.

stability: Atmospheric turbulence; a function of vertical distribution of atmospheric temperature.

stability class: Term used to classify the degree of turbulence in the atmosphere.

stable air: Air with little or no tendency to rise, usually accompanied by clear, dry weather.

stable atmosphere: Marked by air that is cooler at the ground than aloft, by low wind speeds, and consequently, by a low degree of turbulence.

standard atmosphere: A hypothetical vertical distribution of the atmospheric temperature, pressure, and density, which by international agreement is considered to be representative of the atmosphere for pressure-altimeter calibrations and other purposes.

standard temperature and pressure (STP): As the density of gases depends on temperature and pressure, defining the pressure and temperature against which the volume of gases are measured is customary. The normal reference point is standard temperature and pressure, –0°C at a standard atmosphere of 760 millimeters of mercury. All gas volumes are referred to these standard conditions.

standing lenticular cloud: A more or less isolated cloud with sharp outlines that is generally in the form of a smooth lens or almond. These clouds often form on the lee side of, and generally parallel to, mountain ranges. Depending on their height above the surface, they may be reported as stratocumulus standing lenticular clouds (SCSLs); altocumulus standing lenticular clouds (ACSLs); or cirrocumulus standing lenticular clouds (CCSLs).

statement: Provides the public with information concerning the status of existing warnings.

station identifier: A group of four alphabetic characters used to identify a location that makes weather observations.

station pressure: The pressure that is read from a barometer but is not adjusted to sea level.

stationary front: The boundary between cool and warm air masses that are not moving.

stationary sources: Source of air pollution emanating from any fixed or stationary point.

stationary wave: Wave (flow pattern with periodicity in time and/or space) that is fixed relative to earth.

steam fog: Fog that is formed when water vapor is added to air that is much colder than the vapor's source. This is most common when very cold air drifts across relatively warm water.

steering winds (steering currents): A prevailing synoptic-scale flow that governs the movement of smaller features embedded within it.

storm: In marine usage, winds 48 knots (55 mph) or greater.

storm relative: Measured relative to a moving thunderstorm, usually referring to winds, wind shear, or helicity.

storm scale: Referring to weather systems with sizes on the order of individual thunderstorms.

storm surge: A rise of the sea level along the shore that builds up as a storm (usually a hurricane) moves over the water. It is a result of the winds of the storm and low atmospheric pressures.

storm track: The path that a low-pressure area follows.

storm warning: A marine wind warning for sustained winds greater than 48 knots (55 mph) or more from a non-tropical system.

straight-line winds: Thunderstorm winds most often found with the gust front. They originate from downdrafts and can cause damage that occurs in a "straight line," as opposed to tornadic wind damage that has circular characteristics.

stratiform: Having extensive horizontal development, as opposed to the more vertical development characteristic of convection. Stratiform clouds cover large areas but show relatively little vertical development.

stratocumulus: Low-level clouds existing in a relatively flat layer but having individual elements. Elements often are arranged in rows, bands, or waves.

stratosphere: A region of the atmosphere based on temperature; it is between approximately 10 and 35 miles in altitude.

stratus: A flat, low, generally gray cloud layer with a fairly uniform base. Stratus may appear in the form of ragged patches but otherwise does not exhibit individual cloud elements as do cumulus and stratocumulus clouds.

striations: Grooves or channels in cloud formations, arranged parallel to the flow of air and, therefore, depicting the airflow relative to the parent cloud.

subadiabatic: The ambient lapse rate when it is less than the dry adiabatic lapse rate.

sublimation: The change from ice directly to water vapor or from water vapor to ice without going through the liquid water phase.

subsidence: Downward moving (sinking) air over a broad area that is associated with warming air and little cloud formation.

subsidence inversion: A type of inversion usually associated with a high pressure system, known as anticyclones, that may significantly affect the dispersion of pollutants over large regions.

subtropical jet: The branch of the jet stream that is found in the lower latitudes.

subtropical storm: A low-pressure system that develops in subtropical waters (north of 20° north latitude) and initially has non-tropic features but does have some element of a tropical cyclone's cloud structure (located close to the center rather than away from the center of circulation).

suction vortex (sometimes suction spot): A small but very intense vortex within a tornado circulation. Several suction vortices typically are present in a multiple-vortex tornado. Much of the extreme damage associated with violent tornadoes (F4 and F5 on the Fujita scale) is attributed to suction vortices.

sulfur cycle: The natural circulation of sulfur through the environment.

sulfur dioxide: A primary pollutant originating chiefly from the combustion of high-sulfur coals.

sulfurous smog: The haze that develops in the atmosphere when molecules of sulfuric acid accumulate, growing in size as droplets until they become sufficiently large to serve as light scatterers.

superadiabatic: The lapse rate when a parcel of air starting at 1,000 meters at 20°C, for example, starts moving downward and becomes cooler and denser than its surroundings. Because the ambient air is unstable, it continues to sink.

supercell thunderstorm: A severe thunderstorm whose updrafts and downdrafts are in near balance, allowing the storm to maintain itself for several hours. Supercells often produce large hail and tornadoes.

supercooled water: Water that stays in liquid form if undisturbed even though it has been cooled to a temperature below its normal freezing point.

supersaturation: The condition that occurs in the atmosphere when the relative humidity is greater than 100%.

surface hoar: The deposition (sublimation) of ice crystals on a surface that occurs when the temperature of the surface is colder than the air above and colder than the frost point of that air.

surface pressure: The pressure that is read from a barometer but is not adjusted to sea level.

sustained winds: The wind speed obtained by averaging the observed values over a one-minute period.

SWEAT index: Severe Weather ThrEAT index. A stability index developed by the Air Force that incorporates instability, wind shear, and wind speeds.

synoptic chart: Chart showing meteorological conditions over a region at a given time; weather map.

synoptic scale (large scale): Size scale generally referring to weather systems with horizontal dimensions of several hundred miles or more. Most high- and low-pressure areas seen on weather maps are synoptic-scale systems.

TAF: A weather forecast of aircraft operations at an airport.

tail cloud: A low, tail-shaped cloud extending outward from the northern quadrant of a wall cloud. Motions in the tail cloud are toward the wall cloud with a rapid updraft at the junction of tail and wall cloud. This horizontal cloud is not a funnel or tornado.

tail-end Charlie: The thunderstorm at the southernmost end of a squall line or other line or band of thunderstorms.

teleconnection: A strong statistical relationship between weather in different parts of the globe. For example, there appears to be a teleconnection between the tropics and North America during El Niño.

temperate zone: The area of the globe between the tropics and the polar regions.

temperature: A measure of the average kinetic energy of the molecules.

temperature inversion: A condition characterized by an inverted lapse rate.

terrestrial radiation: The total infrared radiation emitted by the earth.

thermal: A small rising parcel of warm air produced when the earth's surface is unevenly heated.

thermal circulation: The result of the relationship based on a law of physics whereby the pressure and volume of a gas is directly related to its temperature.

thermal inversion: A layer of cool air trapped under a layer of less dense warm air, thus preventing reversing to the normal situation.

thermal NOx: Created when nitrogen and oxygen in the combustion air (e.g., within an internal combustion engine) are heated to a high enough temperature (above 1,000K) to cause nitrogen (N_2) and oxygen (O_2) in the air to combine.

thermal pollution: Increase in water temperature with harmful ecological effects on aquatic ecosystems.

thermal radiation: Heat energy directly radiated into space from the earth's surface and atmosphere.

thermodynamics: In general, the relationships between heat and other properties (such as temperature, pressure, density, etc.). In forecast discussions, thermodynamics usually refers to the distribution of temperature and moisture (both vertical and horizontal) as related to the diagnosis of atmospheric instability.

thermometer: An instrument for measuring temperature.

thermosphere: A region of the atmosphere based on temperature between approximately 60 to several hundred miles in altitude.

Theta-e (or equivalent potential temperature): The temperature a parcel of air would have if (1) it was lifted until it became saturated, (2) all water vapor was condensed out, and (3) it was returned adiabatically (i.e., without transfer of heat or mass) to a pressure of 1,000 millibars.

thunder: The sound caused by a lightning stroke as it heats the air and causes it to rapidly expand.

thunderstorm: A storm with lightning and thunder, produced by a cumulonimbus cloud, usually producing gusty winds, heavy rain, and sometimes hail.

tilted storm or tilted updraft: A thunderstorm or cloud tower that is not purely vertical but instead exhibits a slanted or tilted character. It is a sign of vertical wind shear, a favorable condition for severe storm development.

topography: Generally, the layout of the major natural and man-made physical features of the earth's surface. Bridges, highways, trees, rivers, and fields are all components that make up this topography.

tornadic activity: The occurrence or disappearance of tornadoes, funnel clouds, or waterspouts.

tornado: A violent rotating column of air, in contact with the ground, pendant from a cumulonimbus cloud. A tornado does not require the visible presence of a funnel cloud. It has a typical width of tens to hundreds of meters and a lifespan of minutes to hours.

Tornado Alley: The area of the United States in which tornadoes are most frequent. It encompasses the great lowland areas of the Mississippi and the Ohio and lower Missouri River valleys. Although no state is entirely free of tornadoes, they are most frequent in the plains area between the Rocky Mountains and Appalachians.

tornado family: A series of tornadoes produced by a single supercell, resulting in damage-path segments along the same general line.

tornado warning: Issued when there is likelihood of a tornado within the given area based on radar or actual sighting. It is usually accompanied by conditions indicated for a severe thunderstorm warning.

total-totals index: A stability index and severe weather forecast tool, equal to the temperature at 850 millibars plus the dew point at 850 millibars, minus twice the temperature at 500 millibars.

towering cumulus: A large cumulus cloud with great vertical development, usually with a cauliflower-like appearance but lacking the characteristic anvil-shaped top of a cumulonimbus cloud.

trade winds: Persistent tropical winds that blow from the subtropical high-pressure centers toward the equatorial low. They blow northeasterly in the northern hemisphere.

transformation: Chemical transformations that take place in the atmosphere (e.g., the conversion of the original pollutant to a secondary pollutant such as ozone).

transverse bands: Bands of clouds oriented perpendicular to the flow in which they are embedded. They are often best seen on satellite photographs. When observed at high levels (i.e., in cirrus formations), they may indicate severe or extreme turbulence.

transverse rolls: Elongated low-level clouds, arranged in parallel bands and aligned parallel to the low-level winds, but perpendicular to the mid-level flow.

triple point: The intersection point between two boundaries (dry line, outflow boundary, cold front, warm front, etc.), often a focus of thunderstorm development.

tropical air: An air mass that has warm temperatures and high humidities and develops over tropical or subtropical areas.

tropical depression: Tropical mass of thunderstorms with a cyclonic wind circulation and winds near the surface between 23 mph and 39 mph.

tropical disturbance: An organized mass of thunderstorms in the tropics that lasts for more than 24 hours, has a slight cyclonic circulation, and has winds less than 23 mph.

tropical storm: An organized low-pressure system in the tropics with wind speeds between 38 and 74 mph.

tropical storm warning: A warning issued when sustained winds of 39 to 73 mph (34 to 63 knots) are expected within 24 hours.

tropical wave: A kink or bend in the normally straight flow of surface air in the tropics that forms a low-pressure trough, or pressure boundary, and showers and thunderstorms; can develop into a tropical cyclone.

tropics: The area of the globe from latitudes 23.5° north to 23.5° south.

tropopause: The boundary between the troposphere and the stratosphere. It is usually characterized by an abrupt change in temperature with height from positive (decreasing temperature with height) to neutral or negative (temperature constant or increasing with height).

troposphere: A region of the atmosphere based on temperature difference between the earth's surface and 10 miles in altitude.

trough: An elongated area of relatively low atmospheric pressure, surface or aloft. Usually not associated with a closed circulation and thus used to distinguish from a closed low; the opposite of ridge.

turbulence: (1) Uncoordinated movements and a state of continuous change in liquids and gases; (2) one of the three Ts of combustion.

turkey tower: A narrow individual cloud tower that develops and falls apart rapidly.

TVS: Tornadic vortex signature. Doppler radar signature in the radial velocity field indicating intense, concentrated rotations—more so than a mesocyclone.

twister: A colloquial term for a tornado.

typhoon: A hurricane that forms in the Western Pacific Ocean.

UKMET: United Kingdom forecast model.

ultraviolet radiation: The energy range just beyond the violet end of the visible spectrum. Although ultraviolet radiation constitutes only about 5% of the total energy emitted from the sun, it is the major energy source for the stratosphere and mesosphere, playing a dominant role in both energy balance and chemical composition.

unstable air: Air that rises easily and can form clouds and rain.

unstable atmosphere: Characterized by a high degree of turbulence.

updraft: A small-scale current or rising air. This is often associated with cumulus and cumulonimbus clouds.

upper-level system: A general term for any large-scale or mesoscale disturbance capable of producing upward motion (lift) in the middle or upper parts of the atmosphere.

upslope flow: Air that flows toward higher terrain and hence is forced to rise. The added lift often results in widespread low cloudiness and stratiform precipitation if the air is stable, or an increased chance of thunderstorm development if the air is unstable

upstream: Toward the source of the flow, or located in the area from which the flow is coming.

UTC: Coordinated Universal Time. The time in the 0° meridian time zone.

UVI: Ultraviolet Index.

UVV: Upward vertical velocity.

valley breeze: System of winds that blow uphill during the day.

valley winds: At valley floor level, slope winds transform into valley winds that flow down-valley, often with the flow of a river.

vapor pressure: The pressure exerted by water vapor molecules in a given volume of air.

variable ceiling: A ceiling of less than 3,000 feet that rapidly increases or decreases in height by established criteria during the period of observations.

veering wind: Wind that changes in a clockwise direction with time at a given location (e.g., from southerly to westerly), or that changes direction in a clockwise sense with height (e.g., southeasterly at the surface turning to southwesterly a lot). Veering winds with height are indicative of warm air advection (WAA).

venting: In pollution control technology, a method of remediating hydrocarbon (gasoline) spills or leaks from underground storage tanks (USTs).

venturi: A short tube with a constricted throat used to determine fluid pressures and velocities by measurement of differential pressures generated at the throat as a fluid traverses the tube.

vertical shear: The rate of change of wind speed or direction, with a given change in height.

vertically stacked system: A low-pressure system, usually a closed low or cutoff low, that is not tilted with height—that is, located similarly at all levels of the atmosphere.

vicinity: A proximity qualifier used to indicate weather phenomena observed between 5 and 10 statute miles of the usual point of observation but not at the station.

VIL: Vertically integrated liquid water. A property computed by RADAP II and WSR-88D units that takes into account the three-dimensional reflectivity of an echo. The maximum VIL of a storm is useful in determining its potential severity, especially in terms of maximum hail size.

virga: Precipitation falling from the base of a cloud and evaporating before it reaches the ground.

virtual temperature: The temperature a parcel of air would have if the moisture in it were removed and its specific heat added to the parcel.

visibility: The greatest horizontal distance in which an observer can see and identify a prominent object.

volatile: When a substance (usually a liquid) evaporates at ordinary temperatures if exposed to the air.

volatilization: When a solid or liquid substance passes into the vapor state.

volcanic ash: Fine particles of rock powder that originate from a volcano and that may remain suspended in the atmosphere for long periods.

volume: Surface area times (x) a third dimension.

vort max: Short for vorticity maximum; a center, or maximum, in the vorticity field of an air mass.

vortex: An atmospheric feature that tends to rotate. It has vorticity and usually has closed streamlines.

vorticity: A measure of the local rotation in a fluid flow. In weather analysis and forecasting, it usually refers to the vertical component or rotation (i.e., rotation about a vertical axis) and is used most often in reference to synoptic-scale or mesoscale weather systems. By convention, positive values indicate cyclonic rotation.

WAA: Warm air advection.

walker cell: A zonal circulation of the atmosphere confined to equatorial regions and driven principally by the oceanic temperature gradient. In the Pacific, air flows westward from the colder, eastern area to the warm, western ocean, where it acquires warmth and moisture and subsequently rises. A return flow aloft and subsidence over the eastern ocean complete the cell.

wall cloud: A local and often abrupt lowering of a rain-free cumulonimbus base into a low-hanging accessory cloud, from 1 to 4 miles in diameter. The wall cloud is usually situated in the southwest portion of the storm below an intense updraft marked by the main cumulonimbus cloud and associated with a very strong or severe thunderstorm. When seen from several miles away, many wall clouds exhibit rapid upward motion and rotation in the same sense as a tornado, except with considerably lower speed. A rotating wall cloud usually develops before tornadoes or funnel clouds by a time that can range from a few minutes up to possibly an hour.

warm advection: Transport of warm air into an area by horizontal winds. Low-level warm advection sometimes is referred to (erroneously) as overrunning.

warm front: Marks the advance of a warm air mass as it rises up over a cold one.

warning: Forecast issued when a particular weather or flood hazard is "imminent" or already occurring (e.g., tornado warning, flash flood warning). A warning is used for conditions posing a threat to life or property.

warning stage: The level of a river or stream that may cause minor flooding and at which concerned interests should take action.

watch: Forecast issued well in advance to alert the public of the possibility for a particular weather-related hazard (e.g., tornado watch, flash flood watch). The occurrence, location, and timing may still be uncertain.

water: A transparent, odorless, tasteless liquid; composed of hydrogen and oxygen.

water box (or box): A severe thunderstorm or tornado watch.

water equivalent: The liquid content of solid precipitation that has accumulated on the ground (snow depth). The accumulation may consist of snow, ice formed by freezing precipitation, freezing liquid precipitation, or ice formed by the refreezing of melted snow.

waterspout: A rapidly rotating column of air extending from a cumulonimbus cloud with a circulation that reaches the surface of the water (i.e., a tornado over water).

water vapor: Water substance in a gaseous state that comprises one of the most important of all the constituents of the atmosphere.

wave: In meteorology any pattern identifiable on a weather map that has a cyclic pattern, or a small cyclonic circulation in the early stages of development that moves along a cold front.

wave crest: The highest point in a wave.

wave trough: The lowest point in a wave.

wavelength: Physical distance of one period (wave repeat).

weather: The day-to-day pattern of precipitation, temperature, wind, barometric pressure, and humidity.

weather balloon: Large balloon filled with helium or hydrogen that carries a radiosonde (weather instrument) aloft to measure temperature, pressure, and humidity as the balloon rises through the air. It is attached to a small parachute so that when the balloon inevitably breaks, the radiosonde doesn't dangerously hurtle back to earth too quickly.

weather synopsis: A description of weather patterns affecting a large area.

weathering: The chemical and mechanical breakdown of rocks and minerals under the action of atmospheric agencies.

wedge (or wedge tornado): A large tornado with a condensation funnel that is at least as wide (horizontally) at the ground as it is tall (vertically from the ground to cloud base).

weight: The force exerted upon any object by gravity.

WFO: Weather Forecast Office. The southern New England WFO is located in Taunton, Massachusetts. Other WFOs for the northeast are located in Albany, New York; Gray, Maine; and Upton, New York.

whiteout: A condition caused by falling and/or blowing snow that reduces visibility to nothing or zero miles, typically only a few feet. Whiteouts can occur rapidly, often blinding motorists and creating chain-reaction crashes involving multiple vehicles. Whiteouts are most frequent during blizzards.

wind: Horizontal air motion.

wind advisory: Issued for sustained winds 31 to 39 mph of at least one hour or any gusts 46 to 57 mph. However, winds of this magnitude occurring over an area that frequently experiences such winds would not require the issuance of a wind advisory.

wind aloft: The wind speeds and wind directions at various levels in the atmosphere above the area of surface.

wind and breezes: Local conditions caused by the circulating movement of warm and cold air (convection) and differences in heating.

wind chill: The additional cooling effect resulting from wind blowing on bare skin. The wind chill is based on the rate of heat loss from exposed skin caused by the combined effects of wind and cold. The (equivalent) wind chill temperature is the temperature the body "feels" for a certain combination of wind and air temperatures.

wind chill advisory: Issued when the wind chill index is expected to be between –25°F and –39°F for at least three hours. This is using the wind chill of the sustained wind, not gusts.

wind chill factor: The apparent temperature that describes the cooling effect on exposed skin by the combination of temperature and wind, expressed as the loss of body heat. Increased wind speed will accelerate the loss of body heat.

wind chill warning: Issued when life-threatening wind chills of –40°F or colder are expected for at least three hours. This is using the wind chill of the sustained wind, not gusts.

wind direction: The direction from which the wind is blowing.

wind shear: Variation in wind speed and/or direction over a short distance. Shear usually refers to vertical wind shear—that is, the change in wind with height; but the term also is used in Doppler radar to describe changes in radial velocity over short horizontal distances.

wind speed: The rate at which air is moving horizontally past a given point. It may be a two-minute average speed (reported as wind speed) or an instantaneous speed (reported as a peak wind speed, or gust).

wind vane: An instrument that determines the direction from which a wind is blowing.

wind wave: A wave that is caused by the action of wind on the surface of water.

windward: Upwind, or the direction from which the wind is blowing; the opposite of leeward.

winter storm: A heavy snow event; a snow accumulation of more than 6 inches in 12 hours or more than 12 inches in 24 hours.

winter storm warning: Issued when 7 or more inches of snow or sleet is expected in the next 24 hours, or 0.5 inch or more of accretion of freezing rain is expected. A warning is used for winter weather conditions posing a threat to life and property.

winter storm watch: A significant winter storm may affect your area, but its occurrence, location, and timing are still uncertain. A winter storm watch is issued to provide 12 to 36 hours' notice of the possibility of severe winter weather. A watch will often be issued when neither the path of a developing winter storm nor the consequences of the weather event are as yet well defined. Ideally, the winter storm watch will eventually be upgraded to a warning when the nature and location of the developing weather event become more apparent. A winter storm watch is intended to provide enough lead time so those who need to set plans in motion can do so.

winter weather advisory: Issued when 4, 5, or 6 inches of snow or sleet is expected in 24 hours; or any accretion of freezing rain or freezing drizzle is expected on road surfaces; or when blowing or drifting snow is expected to occasionally reduce visibility to one-quarter mile or less.

wiresonde: An atmospheric sounding instrument that is used to obtain temperature and humidity information between ground level and height of a few thousand feet; this instrument is supported by a captive balloon while traveling from the ground level.

World Meteorological Organization (WMO): A specialized United Nations agency responsible for the establishment of meteorological stations and networks and the monitoring of meteorological observations.

wrapping gust front: A gust front that wraps around a mesocyclone, cutting off the inflow of warm, moist air to the mesocyclone circulation and resulting in an occluded mesocyclone.

WSR-88D: Weather Surveillance Radar—1988 Doppler; NEXRAD unit.

yellow wind: A strong, cold, dry west wind of eastern Asia that blows across the plains during winter and carries a yellow dust from the desert.

youg: A hot wind during unsettled summer weather in the Mediterranean.

young ice: Newly formed flat, sea, or lake ice generally between 2 and 8 inches thick.

zigzag lightning: Ordinary lightning of a cloud-to-ground discharge that appears to have a single lightning channel.

zodiac: The position of the sun throughout a year as it appears to move through successive star groups or constellations.

zonal flow (zonal wind): Large-scale atmospheric flow in which the east-west component (i.e., latitudinal) is dominant.

zone of maximum precipitation: The belt of elevation at which the annual precipitation is greatest in a mountain region.

zulu time: Same as UTC, Coordinated Universal Time. It is called zulu because Z is often appended to the time to distinguish it from local time.

Index

abiotic, 155
ablation, 155
absolute humidity, 59, 155
absorption, 92, 155
accessory cloud, 155
accretion, 155
accuracy, 150–151
acid, 33, 46, 162, 166, 207,155
acid rain, 156
acidic deposition, 156
additive data, 155
adiabatic, 41, 156, 174, 178, 190
adiabatic lapse rate, 41–42
adsorption, 156
adsorptive site density, 156
advection, 89
advective, 156
advective wind, 129, 156
aerobic, 156, 162
aerobic processes, 156
aerosol, 156
aerovane, 157
air, 41–44, 157
air mass, 157
air currents, 17, 157
air-mass thunderstorm, 157

air masses, 18, 109–112
air moles, 47
air motion, 128
air parcel, 157
air pollutants, 3, 157
air pressure, 23–35
air stripping, 157
air temperature, 29–30
albedo, 87, 90–91, 158
albino rainbow, 116
algac, 158
aliphatic hydrocarbon, 158
altocumulus, 55
altostratus, 55
anabatic, 158
alkalinity, 158, 195
alkanes, 158
alkenes, 158
alkynes, 158
altimeter, 158
anafront, 159
anemometers, 146
aneroid barometer, 24–25
angular momentum, 159
anticyclone, 159
anvil cloud, 159

arcus, 159
argon, 38
arctic high, 159
atmosphere, 44–51
auger, 160

Babcock, W., 4
backdoor cold front, 161
ball lightning, 161
barograms, 161
barometer, 24
bear's cage, 162
Beaufort scale, 27
Bergeron process, 62–63
bias, 151
biogeochemical cycles, 162
black ice, 163
blocking high, 163
boiling point, 81
bomb cyclone, 175
bow echo, 164
Boyle's law, 34
Brantly, J., 9
breezes, 16
butterfly effect, 139–140

calm, 152, 164
calorie, 82
Campbell-Stokes recorder, 31
cap, 164
capacity factor, 136
carbon factor, 165
cellometer, 165
Charles's law, 35–36
Cheng, K. T. C., 11
Chinook wind, 166
chlorofluorocarbons, 101
cirrocumulus, 55
cirrostratus, 55
cirrus, 55–56, 167
clear slot, 167
climate, 15–16, 106–116
climate change, 92–104
cloud burst, 168
cloud classification, 55

cloud formation, 53–54
cloud height, 33–34
cloud types, 54–56
coalescence, 62
cold front, 18, 110, 168
collar cloud, 169
condensation, 50–51, 53–54, 60, 65, 169
conduction, 80, 82, 89
confluence, 170
contrails, 57
convection, 16, 80–81, 170
convectional precipitation, 64
convective clouds, 54
Coriolis force, 130–133, 170
cosmogony, 44
Cressman, G., 11
cumulonimbus, 55
cumulus, 55–56
cup anemometer, 28
cyclone, 171

damping ratio, 152
dart leader, 171
data-loggers, 145
daylight, 87–88
degree day, 172
delay distance, 152
dew point, 60
diamond dust, 172
differential heating, 88
distant constant, 152
dirty ridge, 173
Doppler radar, 9, 11, 14, 150
drizzle, 61
dry adiabat, 174
dyne, 90
Dyrenforth, R. G., 4

Earth's heat balance, 83
eddy, 175
El Chicon, 12
El Niño, 12, 16, 99, 108–109
electric-contact transducers, 145
energy, 78
ENSO, 12

enthalpy, 41
entropy, 176
environmental science, 16
evaporation, 33, 65–66
evapotranspiration, 58–67

fall equinox, 72
fall wind, 177
Fata Morgana, 115
few, 177
foehn, 178
forested microclimate area, 121
farm animals, 2
freshet, 179
freezing precipitation, 62
friction (drag), 133
Friday, E. W., 13–14
front, 18, 110
frontal precipitation, 65
frozen precipitation, 62

gale, 179
Galileo, 30
gas conversion, 38
Galveston, Texas, 5
gas density, 40
gas flow rate, 38
gas laws, 34–38
geologic era, 94
geologic ice age, 103
geologic period, 94
glaze, 180
global air movement, 127
global warming, 100
glory, 114
graupel, 61
grab sample, 180
Grant, U.S., 4
Great Ice Age, 95
Gregg, W. L., 7
Greely, A., 4
greenhouse effect, 86, 100

hail, 61
Hallgren, R., 12

harmattan, 181
Harrington, M. W., 5
Harrison, 4
heat capacity, 41
heat distribution, 88
heat index, 135
heat island, 72–73
heiligenshein, 114
helicity, 182
hygroscopic nuclei, 62
Hoover Dam, 7
human-caused precipitation, 73–74
humidity, 17
Hurricane Agnes, 11
Hurricane Hugo, 13

ice fog, 183
ideal gas law, 36–38
inflow jets, 184
insolation, 86
interglacial, 97
interrupted light beam, 145
ions, 2
isobars, 129
isohyets, 185
isobar, 185

Jefferson, T., 3
jet stream, 134

Keller, H., 2
Kennedy, J. F., 10
kinetic energy, 78
Kittinger, J., 12

Landsberg, H., 8, 13
La Porta anomaly, 74
lapse rate, 42
La Niña, 186
latent heat, 186
latent heat of fusion, 81
latent heat of vaporization, 81
left mover, 187
liquid precipitation, 60
Lindbergh, C. A., 7

Little Ice Age, 96–97
local air circulation, 133–134
long-term global warming, 102

mamma cloud, 188
Marconi Company, 5
Marvin, C. F., 6
melting point, 81
mesonet, 189
meteorological phenomena, 15
meteorological sensors, 153
meteorological tools, 140–141
meteorologist, 8
meteorology, 15–19, 107–108
Meyer, A. J., 4
microclimates, 118–123; types of, 119–123
Milankovitch hypothesis, 97
mist, 61
mixing, 190
mixing height, 149–150
mock sun, 114
moisture, 52–57
mole fraction, 39
Monticello, 3
morning glory, 191
Morton, J. S., 5

net radiometer, 149
nimbostratus, 55
nitrogen cycle, 192
nowcast, 193

occluded front, 110
olfactory spectrum 2
omega, 193
open land microclimate, 120–121
optical phenomena, 113
orographic precipitation, 64
outflow, 194
ozone, 49
ozone hole, 98

partial pressure, 39
particulate matter, 50–51

Pascal's Law, 34
photosynthesis, 46
plume, 195
pollution, 33
potential energy, 78
precipitation, 25–27, 58–67
precision, 151
pressure, 196
pressure gradient force, 129
profiler, 197
propeller anemometer, 144
psychrometer, 31
pulse storm, 197
pyranometer, 149

quality control, 153–154

radiation, 77–104, 148
radiation budget, 83–84
rain, 61
rain gauge, 25
rain forest, 198
range, 152
reflectivity, 102
refraction, 113
Reichelderfer, F. W., 8
relative humidity, 17, 31–32
reradiation, 84–86
resistance temperature detector, 147
return flow, 199
rime, 61
Rogers, C. P., 6
Roosevelt, F. D., 7–8
rope, 200
rotating cup anemometer, 144

scattering, 91
seaside microclimate, 122
seasons, 68–75
shear, 202
seed clouds, 201
sleet, 61
Smithsonian Institution, 4
smog, 203
snow, 61

snow gauges, 27
SODAR, 150
solar constant, 86
solar energy, 32–33
southern oscillation, 108–109
sounding, 204
specific heat, 81
spring equinox, 70
stable air, 205
steam fog, 206
Stoll, E. H., 12
STP, 23
stratocumulus, 55
stratosphere, 49
stratus, 55–56
sulfur cycle, 207
summer solstice, 70
sun's rays, 88
synoptic chart, 208
system accuracy, 151

temperature difference, 148
thermal inversion, 18
thermal properties, 80
thermistors, 147
thermometers, 30–31
threshold, 152
tilted storm, 209
time constant, 152
Torricelli, E., 24
transducers, 146
transferring energy, 81–82
transparency, 87
transpiration, 66
transport of heat, 89
triple point, 210
troposphere, 48, 149–150
types of winds, 127

urban area microclimate, 122
U.S. National Weather Service, 3–15

valley breeze, 211
valley micro-climate, 121
Van Allen radiation belt, 10
vane-oriented anemometer, 29
vertical shear, 212
viscosity, 42–43
volatile organic compounds, 101
volume percent, 39

walker cell, 213
warm front, 18, 110
Washington, G., 3
water balance (U.S), 66
water vapor, 57
weather, 15–16, 106–116
weather map, 6
weather prediction, 139–154
weather symbols, 8
weather vane, 28
White, R. M., 10
wind, 16, 27–29
wind chill, 134
wind energy, 136
wind forms, 136–137
wind measuring, 146–147
wind shear, 215
wind speed, 141–144
wind speed transducers, 144
wind turbine bird kill, 137–138
wind vanes, 145–146
winter, 71–72
whiteout, 214

yellow wind, 216

About the Author

Frank R. Spellman is a retired U.S. naval officer with 26 years of active duty, a retired environmental safety and health manager for a large wastewater sanitation district in Virginia, and a retired assistant professor of environmental health at Old Dominion University, Norfolk, Virginia. He is the author or co-author of 78 books and consults on environmental matters with the U.S. Department of Justice and various law firms and environmental entities across the world. He holds a BA in public administration, a BS in business management, and an MBA and PhD in environmental engineering. In 2011, he traced and documented the ancient water distribution system at Machu Pichu, Peru, and surveyed several drinking water resources in Coco and Amazonia, Ecuador.